FOREST NURSERY

— HOW TO RAISE AND MANAGE —

FOREST NURSERY

— HOW TO RAISE AND MANAGE —

Dr. S.S. Sagwal

Ph.D. (Forestry), MBA (HRD & MM)
Professor Forestry (Retd.).
S.K. University of Agricultural Sciences and Technology (K). Shalimar (Srinagar), Jammu and Kashmir State (India)

SCIENTIFIC PUBLISHERS

Published by:

SCIENTIFIC PUBLISHERS (INDIA)
5A, New Pali Road, P.O. Box 91
Jodhpur 342 001, India
E-mail: info@scientificpub.com
Website: www.scientificpub.com

ISBN: 978-93-88043-99-1
eISBN: 978-93-88043-90-8

Printed in India

ACKNOWLEDGEMENTS

I acknowledge all the sources of information on the present state-of-out of great interest. I am really grateful to all the friends and colleagues and the critics of present work. Also, I am indebted to m/s Kalyani Publishers for their prompt action in bringing out this small but esteemed, work inspite of their busy schedule. I am thankful to may elder daughter Dr. Poonam Sagwal (MD Medicine) and my younger daughter Dr. Alpana Sagwal (MBA IT & Marketing) and her husband Dr. Rohit Dhakla (MBA Finance from USA) for their endless affection. I am equally thankful to my lovely son Dr. Kuldeep Sagwal, Manager (Syndicate Bank) and daughter-in-law Mrs. Prabhat Tamak Sagwal, Assh. Professor (ECE), who remained very helpful during the writing course of this book. My lovely little master Paarth is very special for me who is a good source of happiness in my life, last but not the least, heart full thanks goes to my wife, Savitri Sagwal (Goddi) for her active cooperation and understanding that enabled me to devote full time to accomplish the writing work of this book.

Kurukshetra, 2018 — Dr. S.S. Sagwal

PREFACE

India is a vast country having diversified flora, fauna and the climatic conditions. The overall development of whole mankind depends upon the natural resources. The plants were the first creatures to appear on this earth. They form the Dias of forests (locally the jungles). However, a severe setback to forests was due to the materialistic nature of man. However, now the government is very much concerned to take care of the green wealth-the forests. To fulfill the cherished aim of expansion of forests there is a dire need to establish very good nurseries of forest plants which can cater the needs of planting stack.

Keeping this in view, the attempt is a little effort in this direction. The book has been written by taking all requirements of an ideal 'forest nursery'. A lot of literature on this aspect has been consulted while taking up this work. Therefore, the author acknowledges fully all the information gathered and compiled systematically. It is hoped that the book will serve a great cause. Suggestions, if any, for further improvement will be solicited.

Kurukshetra, 2018

Dr. S.S. Sagwal

CONTENTS

Chapter 1

INTRODUCTION TO FOREST NURSERY

INTRODUCTION

The main objective of forest nursery is to nurse the forest crop in same way as the woman nurses her baby. In the real sense, the nursery crops (seedlings and saplings) are the essence of the future plantations/afforestation programmes. Hence, a good forest nursery is very important which lays the foundation of forest/forest crops.

DEFINITION

A good and healthy forest nursery is very essential for afforestation and planting-out programmes of any kind. Following are some of the most accepted definitions of forest nursery as mentioned below:

1. Nursery is, an area in which seedlings are raised to be transplanted elsewhere. (General)
2. BCFT (1953). An area where plants are raised for eventual planting out; has ordinarily both the seedlings and transplant beds.
3. Society of American Forester (1983), An area set aside for raising of young trees, mainly for planting out.

The forest nursery usually comprises of the following components such as:

1. Nursery beds;
2. Paths;
3. Irrigation channels;
4. Manure pits; etc.

However, the nursery bed is defined as, a prepared area in a nursery where seed is sown or into which transplants or cuttings are put. Here, two terms have been used transplants and cuttings. The transplants and cutting are two major important process of forest nursery.

Transplant

A **transplant** is a seedling after it has been moved one or more times in a nursery, in contrast to a seedling planted out direct from seed bed.

Cutting

Cutting is defined, as a short length of stem, or root placed in the soil, or other medium, in order to develop into a plant. The stems or root-cuttings are taken from the plants/

trees and put into moist soils where they strike root at the base and these develop into adventitious roots which grow into new plants.

Practices of Cuttings

Following practices are generally followed in cutting as disccused below:

1. **Lopping:** It is a method of cutting which pertains to cutting of branches. However, it is defined as, the cutting-off one or more branches of tree, whether standing, felled or fallen. The promising examples are, *Grewia optiva* (Biul), *Butea monsoperma* (Shak), *Celtis australis* (Khirak), Ban-oak (*Quercus leucotricophora*), Mulberry (*Morus alba*), etc.
2. **Pruning:** SAF (1983), defined pruning in the following two ways.
   (a) **Generally**, considered cutting away of superfluous growth (including the roots) from any plant, so as to improve its development, fruitfulness, etc.
   (b) **Specifically**, considered removal, close to or flush with the stem of side branches, live or dead, and of multiple leaders from a standing, commonly plantation, the grown tree for the improvement of trees or its timber, i.e., stem pruning. However, depending upon the conditions of tree species (living or dead) the pruning is of types, viz.,
   - **Green pruning:** The removal of live/green branches.
   - **Dry pruning:** The removal of dead/dry branches.

   Some other types of pruning are also recognized, such as:
   - Brushing pruning;
   - High pruning;
   - Selective high pruning.

OBJECTIVES AND IMPORTANCE

Forest nursery plays critical role in the afforestation and artificial regeneration. The importance has been increased manifold because of following points such as:

1. To raise fast-growing species;
2. Short-rotation crops;
3. Introduction of suitable exotics.

The main objectives for which forest nursery is raised, adds the crown to its importance:

1. It has been seen by experience that some tree species do not seed every year. Therefore, plantation of such tree species should be raised annually by obtaining the seed from such species where the seal is available in sufficient quantities in the years of moderate and good seed years. The sowing of such seed is carried out in the nursery for growing/raising seedlings for planting out in various years.

2. The success of roadside avenue plantations, usually depends on tall, beautiful (plus trees), and sound and sturdy plants/trees, which are supposed to obtain from the nursery only.
3. There are evidences that many tree species grow with slow pace. If, the seeds of such species are sown directly in plantations, the weed infestation may kill such seedlings which is a setback to the stock of plants. Here, the slow-growing species must be raised in nurseries and when these are of out of reach the attack of weeds, then such plants must be planted out with a great care by avoiding the damages in such cases.
4. It has been noticed that the direct sowing plantations are not so successful as when these are raised/grown by direct sowing. Therefore, raising of nursery is supposed to be necessary and essential part of the afforestation of such species.
5. The sure set method of introduction of exotic species, viz., some species of primes, eucalyptus, spp., populus spp., is only by planting out and thus, nursery is very important in such cases.
6. Poor and barren sites are the ideal areas for planting out nursery-grown plants seedlings/and sometimes, the saplings also.
7. The failures in plantation areas are to be replaced either in the year of formation/ or in the next year. The sowings which are carried in the vacant areas/gaps are usually liable to be unsuccessful due to the infestations of the weeds. This causes a big gap between the newly planted and the original crops/sowings.

Therefore, the objectives are well-set and must be fulfilled by raising the forest nurseries so that a ground success is easily achieved.

CLASSIFICATION

The nurseries are classified in several ways, viz.,

1. **Based on irrigation facilities:** The irrigation facilities decide the fate of nurseries. These are of following types:

   (a) Dry nursery: Refers to, the nursery which is maintained without any irrigation or other artificial means of watering.

   (b) Wet nursery: It is, a nursery that is maintained by irrigation or other artificial watering during the dry periods.

2. **Based on nature of beds:** The nursery beds are classified, on the basis of the kind of plants growing in the nurseries. These nurseries are classified as:

   (a) Seedling nursery: A nursery which has only seedling beds, i.e., in which the seedlings are raised, no transplanting being done.

   (b) Transplant nursery: A nursery which has only transplant beds, i.e., in which seedlings are transplanted in preparation for forest planting.

   However, in general, whatever is raised in a nursery for planting out is known as nursery stock. Also, this term is used for plants supplied from a nursery. In

India, it is a practice, that separate seedling and transplant nurseries are not made. However in the same nursery, some beds are used as seedling beds and others as transplant beds.

3. **Based on use:** Here, following two types of nurseries are dealt with:

(a) **Temporary nursery:** These are also referred to as, field nursery. It is defined as, a temporary nursery formed in or adjoining the planting area/site. These are raised for supply of forest nursery stock for a short period and these are established for producing seedlings for 2–4 years and does not have permanent structures (e.g., roads, and buildings).

Advantages

- These are made in the areas of newly cleared and are rich in humus/organic matter.
- There is minimum trouble with the weeds, inject-pests and diseases.
- It also enables easy, quick and cheap transport of transplanting stock without any damage.

Disadvantages

- Here, the proper supervision is not possible due to the out-of-way places.
- Due to the absence of irrigation facility, the growth of seedlings is somewhat slow and a heavy mortality occurs.
- This type of nursery is costlier to maintain in comparison to permanent nursery.

(b) **Permanent nursery:** This type of nursery is maintained for supplying nursery stock for a long span of time on a permanent basis. The duration of service-life is very long. This type of nursery is generally made in a central place in a range in the vicinity of headquarter. Its fertility reduces due to the long time of raising the plants in such nurseries.

SALIENT FEATURES

These nurseries are kept for longer period of time and passes some salient features which are as follows:

1. These types of nurseries are run and maintained for longer period making them as permanent;
2. These nurseries are bigger is size and extent in comparison to the temporary nurseries;
3. These are maintained for a larger variety and the number of tree species;
4. These nurseries are the permanent source of irrigation, buildings, pathways, and some other assets, and;
5. In such nurseries, a large number of labour forces is required.

EXTENSION NURSERIES

These are also called decentralized nurseries. These are not directly manage by an official agency. These types of nurseries may comprise of the following types:

1. **Kisan/Farmer's nursery:** Sometimes, the farmers/kisans cannot get the seedlings/saplings of the desired tree species from some reliable sources and, if, they get them that too on some unbearable cost which is beyond the capacity of the kisans/farmers. In such conditions, the farmers go for establishing their own nurseries. The advantages of such nurseries are,
   (a) These are within the range of the farmer's economy status;
   (b) The desired species may be grown in such nurseries;
   (c) These are less costly and the farmers can handle them with great efficiency.
2. **School nursery:** Generally, the schools may also establish the nurseries at their own which are the immediate source of seedlings and saplings. The schools can distribute the planting stock to the nearby areas in the villages also. Even the students may also be assigned the duties to look-after these types of nurseries. This is good habit and the moral character of the student may be enhanced. They may be responsible citizen at later stages after their schooling. Therefore, such types of nurseries should be undertaken.
3. **Village nursery:** In India, every year the festival of trees (Vana Mahotsava) is celebrated with great interest and a large number of people take part in this venture. The 'Panchayat Ghars' in the villages can be the most ideal place for establishing the nurseries. Therefore, depending upon the local conditions, the desired tree species can be grown in such nurseries. The advantages of such nurseries are:
   (a) These are less costly.
   (b) The land of Panchayat Ghars can be used for this purpose.
   (c) The villagers can take part in such activities and may make it a ground success.

Therefore, the extension types of nurseries are valuable as the local resources can be utilized properly and it will be beneficial exercise.

Chapter 2

SELECTION OF SITE

INTRODUCTION

The nursery site selection is a very crucial factor in the establishment of nursery. It should be done with considerable care and caution. As a general rule, minimum requirements and needs, each nursery site should have the following:

1. Sufficient area;
2. Suitable climate and soils;
3. Adequate energy;
4. Transportation facilities;
5. What of sufficient:
   (a) Quantity; and
   (b) Quality
6. Sufficient labour force for various works.

IMPORTANT SELECTION FACTORS

The selection of site is a very important step in nursery. The most important factors, which are taken into consideration for site selection are discussed below:

1. **Climate:** The growing-season requirements will vary with stock type. Following are the factors of climate that affect the site selection are below.
   (a) **Temperature:** Extremely hot periods reduce growth and can cause burning of foliage. In the same fashion, extreme cold can derive frost deep into the soil, delaying lifting and the seedlings processing well into spring.
   (b) **Precipitation:** High rainfall must be avoided for nursery site. The summer rains may become a problem only when they occur in brushing of clouds. This may result in a number of problems such as, flooding, erosion, or the seedlings wash-out. Therefore, the scattered and heavy rainfall are detrimental to the growing stock in the nursery.
   (c) **Wind:** The area with heavy frequency of wind must not be chosen for nursery site. The application of irrigation is adversely affected by the winds. The high winds can usually desiccate the growing seedlings.
2. **Soil:** The soil is supposed to be the most crucial factor in establishment of nursery. The soil characteristics are very important in this case. A brief account of the soil factors is given below:

(a) **Texture:** A good nursery requires well-drained soil system. The clay or silt contents should be at minimum level. The preferred soil of nursery should be sandy or loamy sand. However heavy and clayey soils must be avoided. The soil with sufficient organic matter is ideal which can retain water and nutrients.

(b) **Depth of soil:** Wilde (1958) suggested that the top layer of 1.3m soil should be free from any clay pan, hard pan, shade, iron concretions, calcareous substrates, mottled gley layers. However, the depth of soil depends upon the drainage system.

(c) **Soil pH:** The optimum pH values of soil is between 5.0 and 6.0. The soils, however, with lower pH value may have fewer nutrient availability whereas the higher pH encourages the invasion of fungal disease which is fatal to the nursery stock. Therefore, pH value should be at optimum level so that it will be beneficial to the growing stock.

3. **Topography and aspect:** The topography for nursery site must be such that frost hollows and the exposed sites are usually avoided. However, the site of nursery should be on gentle slope that may have sufficient gradient. This will help in allowing the rain-water to run-off without causing erosion problem. However it should be remembered that the flat grounds may become water-logged. For nursery sites, particularly in hilly tract, the northern and coaler aspects should be preferred. However, the easterly aspects are not good for site selection for nursery. This may be due to the reason of morning thawing cold winds, etc. Hence, topography and others very important aspects play role in affecting the selection of site.

4. **Water:** It is usually believe that the supply of irrigation water may pose a great problem. The water (moisture) is very important in the growth and development of the plants at nursery stage and also it protects from the frost. For ensuring the supply of water and the quantity there of, then following points must be considered:

   (a) **Irrigation-water sources:** The lakes, streams, wells, tube-walls, and pipelines, etc., are the ideal sources of irrigation water.

   (b) **Water quality:** The chemical contaminants may be introduced into an irrigation source through the soil or from precipitation, or surface run-off. The water of well may be contaminated with some minerals. Therefore, for selection of site, the evaluation of water of contamination or polluted conditions must be carried out before hand and to start the work of site selection.

5. **Previous land-use:** The past use of may have its effect on the value as a potential nursery site. It should be confirmed whether the site have been altered or not. Therefore, before recommending it for nursery establishments, the past history of the site should be looked into very carefully and the present vegetation must be examined very carefully for the following few factors, viz.,

(a) Root diseases;

(b) Foliar application;

(c) Disorders of any kind; etc.

For site selection, following points must be considered very carefully:

- Whether pesticides ware used or not?
- Are there any known insect-pests problems?
- Is there any evidence of weed infestation?
- What was the source of irrigation?
- Was there any drainage system on the site?
- However, there could be many other queries which must be enumerated and valuated. It is important to see the previous land-use before taking up any programme of establishing the nursery.

6. **Nearness to plantations:** In this connection, mainly two points are very essential, viz.,

(a) **First,** the nursery should be as far as possible near to the plantations/ plantation site; and

(b) **Secondly,** the nursery should be located at road-head.

The nearness of nursery to plantation site will certainly reduce the cost of transportation of planting stock and thus, will save a lot of time.

7. **Availability of labour:** This is an essence of an ideal nursery and therefore, nursery must be within the limit of easy access of labour in the area. The number of men required should generally depend upon the following factors:

(a) The size, capacity and degree of mechanization; and

(b) The general employment opportunity and situation in a particular locality.

In addition to the labour availability, there is an urgent need of trained technical staff so that the supervision may be fruitful. However, the untrained or inexperienced supervisory staff may spoil the interest of establishing the forest nursery.

8. **Other factors:** There are some more/other factors which affect the nursery site. Following are the important factors, viz.,

(a) Site production potential;

(b) Proximity to the following:

- Customer;
- Labours; and
- Service.

(c) Transportation facilities;

(d) Other services such as:

- Garbage disposal system; and
- Drainage system

(e) Land availability in the area/site; and

(f) The cost factor.

In addition to the above factors of selection of site, there are following factors which are also very important such as:

1. **Air:** It has been noticed that air pollution is a potential problem for nurseries which are located in the vicinity of highly industrialized areas and also in the rural areas. Some pollutants like, sulphur dioxide (SO_2), Hydrogen fluoride (HF), and Ozone (O_3) can be carried out easily downward from their source on the nursery site and settle down directly on the seedlings (Davidson 1981). The wind acts as a carrier of some pollutants for long distances. This may cause the deposition on seedlings foliage and it may even cause the death of the seedlings in the nursery site.

2. **Equipments and tools:** Galloway (1981) reported that organization and layout of a nursery require space for, (a) storage; and (b) maintenance of equipments and tools. An inventory system should be implemented from the very starting of nursery operations. For want of proper inventory of equipment and tools, they are either lost/stolen and maybe purchased in duplicate when this cannot be traced out. There is a huge loss of money and time, if, the tools and equipments are not available. Therefore, one person should be given the duty for the following activities:

   (a) Issuance of tools and equipments;

   (b) Retrieval;

   (c) Operation;

   (d) Maintenance;

   (e) Repair of equipments and tools.

It has been suggested that a bimonthly maintenance schedule for important and major equipments should be implemented. Rusting of these equipments may be a very big problem. Therefore, these equipments should carefully be sanded (treatment with sand) and repainted with some good paints should be performed. In all types of climates, preventive maintenance is the only method to give long life to the nursery equipments. Moreover, at the end of planting season, an equipment inventory should be taken immediately.

Following is the probable list of equipments which are frequently used in nursery and its operations:

1. Anemometer.
2. Apron, cloth.
3. Apron, rubber.
4. Auger, soil type.
5. Balance, analytical.
6. Balance, bar.
7. Balance, spring type

8. Basket, metal.
9. Basket, woven.
10. Basket, plastic.
11. Burch, sitting.
12. Boots, rubber.
13. Boxes, wooden bottom.
14. Boxes, wine mesh botton.
15. Buckets, galvanized.
16. Buckets, plastic.
17. Cans, watering.
18. Carpenters suppliers and tools.
19. Cloth bags.
20. Containers, large capacity.
21. Containers, polythene.
22. Container, tim.
23. Crowbar.
24. Cutters, wire
25. Files.
26. Filter paper for seed germination purposes.
27. Fire extinguisher.
28. First aid box.
29. Germinator.
30. Gloves, clothes.
31. Gloves, rubbers.
32. Graduated cylinders.
33. Grafting tapes.
34. Hormones for vegetative propagation.
35. Knives, grafting.
36. Knives, pruning.
37. Laboratory chemicals.
38. Glass wares.
39. Microscope.
40. Mixer, soil.
41. Petri dishes.
42. pH meter, electric and portable.
43. Pickaxe.

44. Plastic trays (for seed germination).
45. Pliers.
46. Polythene bags.
47. Polythene sheeting, black.
48. Rates, garden.
49. Refrigerator, if required (feasible).
50. Rulers, metric.
51. Seed moisture tester.
52. Seed separator.
53. Seed sieve.
54. Shovels.
55. Soil sieves (large size).
56. Spray tanks (for fertilizer and pesticides).
57. Storage bins.
58. Tape, measuring.
59. Test tubes.
60. Thermometers, recoding.
61. Wire, galvanized.
62. Wire mesh (6 mm and 13 mm sizes).
63. Wrenches (plumbing types).

Chapter 3

DESIGN AND LAYOUT

INTRODUCTION

A permanent type of nursery is supposed to be a long-term. Commitment and investment is also the important factor. A good design and layout will increase the efficiency of growing the seedlings. A real design and layout of nursery depend on whether a nursery:

1. Produces bare-root or containerized seedlings.
2. Uses mechanized or non-mechanized techniques of raising nursery; and
3. Has a large or small capacity.

Further, anticipated operation as a commercial ornamental (Davidson 1981) or commercial greenhouse (Tinus and McDonald 1979) is also a great concern in the design and layout of an ideal nursery.

AREA OF NURSERY

The area of nursery generally depends upon many factor such as:

1. Species to be raised;
2. Age of seedlings/transplants;
3. Number of transplanting required;
4. Area of plantations;
5. Spacing in plantation.

However, the area estimated on the basis of above mentioned factors should be increased by 50% for making provision of paths, roads, irrigation channels, etc. However, Davidson and Mecklenburg (1981); Escobar et al. (1975) reported that general considerations are size of entire complex, expected yearly production rates for species grown, seedling density in beds, kinds of equipments used, whether bed orientation is square, or rectangular and parallel or diagonal to major buildings.

Following two conditions are usually taken into consideration as discussed below:

1. **Bare root beds:** The following are taken into consideration, viz.,
   (a) Number of rows in each bed;
   (b) Seedling spacing in each row; and
   (c) Minimum aisle width required to accommodate the equipments.

2. **Container beds:** The specific views are whether:
   (a) Beds are raised or ground level type;
   (b) Have hard or soft surfaces;
   (c) Have wide or narrow benches; and
   (d) Containers used are unitized or individual cells. (Legel and Venator 1987).

Regarding nursery areas under Indian conditions, following are few examples (Khanna 1984):

1. **Uttar Pradesh and Uttarakhand:** It is viewed as under:
   (a) For decoder (*Cedrus deodara*), 30 beds of nursery, of 1.8m × 1.2m are supposed to be sufficient for 1.0 area (0.4 hectare) plantation where 2 years and 3 months old plants are planted at a spacing of 1.8m × 1.8m. Therefore, the area of nursery beds works out to be as 64.8 sq. m and after adding the area occupied by paths, terrace slopes, etc., then area of nursery becomes nearly 100.0 sq. m., i.e., 2.5% of the plantation area.
   (b) For chirpine (*Pinus roxburghii*) grown in polythene bags and planted-out at a spacing of 3m × 3m a nursery of the area of 25.0 sq. m is supposed to be sufficient for one hectare of plantation, i.e., the nursery should be 0.25% of the plantation area.
2. **West Bengal:** For 1.8m × 1.8m planting of one year old nursery stock, one hectare nursery is sufficient for plantation of 200 hectares, i.e., 0.50% of the plantation area.
3. **South India:** The nursery beds 12.2m × 1.2m sown 9 to 13.6 kg seed produce sufficient stamps of plan table size to be planted at 1.8m ×1.8m for a plantation area of 0.4 hectare (one acre). Here, area of nursery bed works out to be 14.64 sq. m. and adding the area of paths, roads, the area of nursery would be about 20.0 sq. m., i.e., 0.5% of the plantation area.

Storage and Operational Areas

The storage facilitates should be large enough to accommodate daily raw material requirements and spare parts for small types of equipments (power saves, mowers, etc.). The operation areas are needed for:

1. extracting, drying, and processing seed;
2. preparing germination trays and pot mix;
3. screening compost;
4. filling pots, and washing, grading, and packaging the seedlings (Weber 1977).

The key concern for these areas is maintaining flexibility and integrated flow of activities (Aldous 1972).

A brief description of flexibility and integrated flow is given as under:

1. **Flexibility:** Refers to the willingness to incorporate new machinery and procedures as required to increase the production or to change the kinds of seedlings grown there in.

2. **Integrated flow:** Pertains to arranging buildings walkways, road-ways, alleys, etc., to complement each other. The space is saved by intentionally designing some areas to accommodate many types of activities.

Herbicide and Insecticide Storage

The chemicals are toxic in nature and, therefore, their storage must be specialized and should be taken care of regularly. The chemicals may include the following:

1. Insecticides;
2. Fungicides; and
3. Herbicides.

All these chemicals should be, at all cost, protected from environmental elements. The suitable storage areas should be:

1. posted with proper signs advising of dangerous areas;
2. kept dry, cool, and well-ventilated;
3. accessed by some authorized personnel;
4. located in aloof areas (buildings) away from other areas used by men and animals; and
5. constructed of fire-proof materials and hard-surfaced shelves and the floors.

Precautions

The chemicals are very dangerous for human-beings. Following precautions must be adhered to:

1. Insecticides and herbicides are always stored separately;
2. All the chemicals should be stored well in their original containers for avoiding confusion with some other products;
3. It is advisable that container lids/caps should be tightly closed to prevent volatilization of liquids and escape of the toxic fumes.
4. For hazardous and dangerous materials, the local authorities may have to authorize final disposal.
5. An up-to-date inventory should be kept of all chemicals, showing remaining amounts of liquid, solid, granular and the gaseous materials.
6. At least one person should be trained in handling the chemicals, their use, and the storage procedures.

Fencing

After the decision of nursery area, and the site has been selected, them a rectangular plot of that area should be demarcated and faced with care. Khanna (1984) gave a good account of fencing the nurseries, viz.,

1. **Temporary nursery:** Located in the previous year's plantation or in some proposed plantations which are already found, need not be found again. However, if, it is located in the unfenced plantation site/area it has to be fence, for example, Eucalyptus hybrid nursery.
2. **Permanent nursery:** This type of nursery is properly fences with cattle proof or game-proof fence. However, this depends upon the extent of damage. A line of small trees/or shrubs, e.g., *Acacia nilotica Duranta plumeri*, etc., about 3 or 4m high may be raised along with fence from the effect of desiccating winds and thus helps the wind erosion problems.

LAYOUT OF NURSERY

There are many points to be considered in laying out the nursery as mentioned below:

1. **Shape of nursery:** The idealistic shape of nursery should be square or near square. This will reduce the length of nursery. It will also minimize the lime in moving from one part to the other. The nursery area should be divided into different sections and the minimum size should not exceed more than 0.4 hectare (1.0 acre). However, it may be more in commercial nurseries.
2. **Leveling land:** After cleaning operation, the field/area of nursery should be leveled and sloped. If, the area/site is already level, with (say) 2 to 3% slope, then adequate finish leveling may be required. The desired slope also depends upon soil characteristics and properties. The sandy soils require less slope than the heavier ones. However, the field/area of nursery can be tiled for supplemental drainage before final leveling and sloping.
3. **Orientation of beds:** The orientation of beds should be considered before final slope and the position of roads/paths are planned will in time. The North-South orientation is important that gives the advantage of receiving the sunlight in full. The East-West orientation, however, may increase the possibility of sun-scaled in summer season.
4. **Providing road system:** Road and paths are the salient features of the permanent nurseries. Metulled roads are always preferred. There should be 5.0 m distance between two sections. The temporary paths within or between the sections should be 1m wide. In the same fashion, there should be provision of buildings for office and store.
5. **Irrigation channels:** After orientation of beds, irrigation system must be designed and installed. Under Indian conditions, the water is given to plants to plants through the channels. These channels may be Kacha or Pucca in construction. There are some instances where one can see the irrigation system as sprinkler irrigation. Though this system is somewhat costly but it is very effective forest nurseries. Thus, irrigation system is the need of the hour for every nursery.

Further, while laying the nursery, the location of irrigation channels should be kept in view. The channels should be made along the paths as far as possible. Moreover, the beds of nursery should be laid out with their lengths east to west. In this way, they can be shaded against the frost and sun without any difficulty. In a very simple notion, there should be a tractor-road on one side of its lengths and an irrigation channel on the other.

PREPARATION OF BEDS

Here two points are usually considered, viz., size of beds; and soil preparation.

1. **Size of beds:** The size of beds varies from locality to locality. There are two options regarding the width of nursery beds, viz.,

   (a) The beds are 1.2m wide; and sometimes it is kept as 1.8 m.

   But in case of hilly areas, the beds are kept 0.9m wide in Deodar (*Cedrus deodar*). However, as a general operation can be carried out easily and that too without any harm to the plants growing therein. Moreover, the length of nursery beds varies from 1.8m in the hilly areas to about 12.2m in the plain areas. By taking the case of West Bengal (India), the nursery size of kept as under:

   Length of nursery beds = 3.6m;

   Width of nursery beds = 1.2m (Sometimes 1.8m).

   These nursery beds are usually called as Kamra. However, in South India, the standard nursery bed is 12.2m × 1.2m (40′ × 4′). But in metric system, the nursery beds of size 10m × 1m are usually recommended.

2. After demarcation, the beds are dug to a depth of 0.3m to 0.45m so that stores, roots, etc., lying underneath are all dug out, picked and thrown outside the nursery.

In wet areas, the nursery beds are generally raised 10–15cm above the level of paths/roads. The earth may be dug out from the paths/roads and then put on the nursery bed to raise the level of beds. Alderex 5%, dust must also be mixed with earth at the rate of 75 kg/hectare. In West Bengal, while mixing of leaf-mould, charcoal dust, and the ashes are mixed thoroughly. It is believed that by adding such mixtures, acidity is controlled and corrected, further, another practice is also followed by burning of dry grass and the strubs put in the beds perform the two tasks, viz.,

(a) Reduces the weed growth; and

(b) Very beneficial for the species with minute seeds, for example, Adina Cord folia

Role of Mycorrhiza

It is believed that when a species which has Mycorrhiza in its roots is supposed to be raised in the nursery bed outside its natural habitat, it is necessary to appropriate mycorrhiza is judiciously introduced. It helps the proper development of the seedlings (Khanna 1984). The mycorrhiza can be introduced by two ways:

1. Inter-planting imported seedlings with the mycorrhizae in the nursery beds; and
2. Mixing the soil of natural good quality forests. The soil should contain in enough quantity of the following:
   (a) Moisture; and
   (b) Mycorrhizal roots.

The soils from the forest areas, should be brought in the sealed polythene bags and put at the desired site at the earliest. However, it should not exceed 10 days of collection of soil.

PROTECTION OF NURSERY

1. **Erecting windbreaks:** The windbreaks are raised on the wind ward sides of production areas and germination beds to reduce the,
   (a) Drying;
   (b) Eroding; and
   (c) Abusive effects of winds on the growing seedlings.

   Kittredge (1973) reported that the ameliorating effects of windbreaks extend many times, the height of mature trees. Moreover, the large nurseries need many lines of windbreaks for protecting the beds.
   (a) As a general rule, the species which are different from those grown in the nursery should be used because it affects the improved seeds.
   (b) The species with large spreading root systems and crowns are usually not suitable. The spreading root systems compete with bare-rooted seedlings for nutrients and the moisture. The outbreaks of insect-pests and diseases must be kept under control.
2. **Shade and bed orientation:** It has been reported elsewhere, that an East/West orientation of nursery beds may be better than the traditional North/South bed orientation. The justification for East/West orientation is usually based on:
   (a) Need for some shade during the hottest part of the day; and
   (b) Reducing the amount of shading material required for providing shade protection for the seedlings.

PRODUCTION SCHEDULING

It is viewed in the following two ways, viz.,

1. General principles; and
2. Specific principles.

Now, a brief dissention of both the principles is given as follows.

1. **General principles:** Legel and Venator (1987) reported that the key to a successful nursery programme is the planning and developing realistic time-

tables for all phases of seedling production is usually called production scheduling. The schedules are based on the knowledge of nursery physical plant limitation and managerial experience.

(a) Each step in a production schedule should be identified and then listed in a sequential manner. The steps of individual nature in the process of production may be converted/translated to a standard work unit, e.g., 1,000 seedlings per labour/man per hour, alldhous (1972) and Davidson (1984) suggested that all nursery operations and works must be constantly under review for potential improvement.

(b) The nursery managers introduce new techniques and refute the ideas and train the nursery supervisors to use them effectively. Moreover, the managers must;

- Make sure labourers use equipment properly efficiently, and safely;
- They must read nursery publications/literature visiting some other nurseries, and attending nursery conventions from time to time.
- Peace (1982) suggested that periodic changes are often required that might upset the work habits and traditions. Therefore, for reducing the conflicts with traditions and personnel, the managers of forests nurseries should educate their own people, who work in the nurseries, about the reasons for and the benefits of any changes.
- The managers must be aware of the particular community's needs and the realities, Weber (1977) and Weidelt (1975) suggested that employing the local labourers promotes community self-help and fosters the grass-roots acceptance of nursery and reforestation objectives.

2. Specific principle:

(a) **Planning and coordination:** The coordination among various individual is very important so the programme of any kind may be made a general success. The coordination is just required so that the owners/managers of forest nurseries donot go for production of the seedlings at the wrong time of the year. Some portion/percentage of each nursery crop may contain seedlings which may not be able to meet out the out planting specifications. In general rule, at least 10% more seedlings should be produced than are required for allowing for cell seedlings. However, this practice is based on the experience in growing/raising forest nurseries.

(b) **Growth rates:** The nursery men rely primarily upon the experience to determine how fast the seedlings of particular tree species grow. For example, if, the seeds of specific tree species are sown in the month of December, it could take 6–7 months to produce good sized seedlings. Therefore, the production scheduling is based on expected growth rates of a particular species. This is, however, determined by,

- Personal knowledge; and
- Previous growth records.

(c) **Disease:** The disease may also kill the seedling to some extent. The disease killing is a high level possibility of seedlings. If, the seedlings are raised and produced on a staggered basis, the danger of total loss of crop from a very single disease is reduced to a great extent.

A skilled nursery man/manager reviews all the phase of production with great care, well in advance. Therefore, planning of production scheduling should be executed with all precautions and thus, the production can be a success story in the forest nurseries.

Chapter

4

SEED COLLECTION AND STORAGE

SEED COLLECTION

The seed collection in forestry in a form of art. If, wrong type of seed is collected of a tree species and ultimately sown, there could very much chances that an inferior crop the forest trees may take place which may not be the cherished aim and objective of the afforestation programmes. In forestry, there has always been a serious problem of seed supply. It is very difficult task to procure adequate quantity of seed from suitable and reliable sources (Khan 1987). Therefore, for smooth functioning of a seed collection programme, following are essential to know:

1. Specialized knowledge and technical skill and know how to certain extent;
2. A good and sound coordination among various operations in valued;
3. Establishing seed centers and seed handling stations; and
4. Various kinds of equipments and tools for the following:
   (a) Cleaning of seed;
   (b) Dewinging of seed;
   (c) Seed grading;
   (d) Dry kiln methods; and
   (e) Other seed extraction methods; and
5. The seed center/stations must have the following facilities:
   (a) Seed testing facilities; and
   (b) Seed storage facilities.

SEED REQUIREMENTS

In forestry, it is very important to known the seed requirements for afforestation programme. Therefore, it is pertinent to determine and calculate the requirement of seed of forest tree species per unit area. The following formula is used for the quantity of seed required for raising a plantation:

$$\therefore \text{ Seed required (kg)} = \frac{A \times D}{S \times N \times G \times P \times M}$$

where, A = Area (in hectare) to be planted in a year for each tree species;

D = Density of plantation (i.e., the number of plants hectare);

S = Survival percentage in the plantation (to be expressed as a decimal);

N = Number of seed per kg (kg^{-1});

G = Germination percentage (as a decimal);

P = Purity per cent of seed (as a decimal); and

M = Mortality percent (%)-as a decimal.

The utility of the above formula can be illustrated as follows:

Species to be planted = Melia azedarach. Lim. Dek, Drek, Bakain (Hindi),

A (Area to be planted) = 100 hectares;

D (Density of plantation) = 2500.

S (Survival %) = 70% (0.70) as decimal;

N (Number of seeds/kg) = 1500.

G (Germination %) = 80% = 0.80 (as decimal);

P (Purity % of seeds) = 90% = 0.90 (as decimal);

M (Mortality %) = 90% = 0.90 (as decimal)

Therefore, By applying formula

$$\text{Seed required (in kg)} = \frac{A \times D}{S \times N \times G \times P \times N}$$

$$= \frac{100 \times 2500}{0.70 \times 1500 \times 0.90 \times 0.90}$$

$$= 367 \text{ kg (say 370 kg)}$$

Therefore, the seed requirement is calculated individually for each tree species which is supposed to be planted under afforestation programme. It is essential and advisable to calculate the requirement of seed before going for seed collection operation.

METHODS OF SEED COLLECTION

There are many methods and equipments available for seed collection and the choice, therefore, depends upon a number of factors (Robbins et al. 1981). Following are the factors affecting seed collection:

1. Collection is done mostly by waiting natural full of seed and collection from the ground. However, on the other end, collection from the tree of fruiting heads of Haldu (*Alina cordilora*) at 200 kg^{-1} is the only practical and behavioral way to collect the seeds but a 11 million kg^{-1}, it would be next to impossible to collect them after the dispersal (Cambell 1980).
2. Characteristics of individual matter a lot. This can, however, be explained as follows:

   (a) Characteristics of fruit: This includes the following:

   - Size;

- Number;
- Position;
- Distribution of seeds;
- Resistance of peduncles to shelling;
- Pulling;
- Breaking or cutting; and
- Interval between: (a) Ripening; and (b) Opening.

(b) **Characteristics of tree:** This usually includes the following:

- Bole: Diameter, shape, and length.
- Bark thickness;
- Shape and depth of crown;
- Density of foliage; and
- Branches, size, angle, density and resistance to breakage.

(c) **Characteristics of stand:** The distribution and staking of trees, for example.

- Isolated trees; and
- Open or dense stand.

(d) **Characteristics of site:** The lays emphasis on the slope accessibility, etc.

CLASSIFICATION OF METHODS

1. Collection of fallen fruits/seeds from the forest floor

(a) **Natural seed fall:** The fruit/seed size is very important as the larger the fruit, the easier will be to see and pick up by hand from the ground. The temperate genera which are collected from the ground include, castanea fagns, queries, etc., and the tropical genera include, gmelina, tectona, triplochiton, etc. The seed of some hard-coated species may remain viable on the forest floor for years, particularly in the temperate conditions, however, the identity of the mother tree is usually uncertain when the fruits/seeds are picked up from the ground.

(b) **Manual shaking:** In some cases, the fruits/seeds full may be induced by some artificial means. The stem of small trees and low branches are shaken directly by hand. However, the higher branches may be shaken by means of a long pole and hook or by a rope. This makes the seed collection easy with good viability when the fruits/seeds are mature (Stead 1979, Robbins et al. 1981).

(c) **Mechanical shaking:** The use of machines for shaking is not a common practice in India. This is just because of the machines are somewhat expensive, need flat ground for efficient use, and experience operators are very essential to avoid excessive destruction and damage to the trees. However, it is a quick method and many cones can be removed in a few seconds, but the longer shaking breaks-off pieces of tops and limbs (stein et al. 1974).

(d) **Collection after dispersal:** The collection can be executed for seed dispersed after the cones/fruits have opened in some tree species; there is very short period between reaching at maturity and its dispersal. Thus, several methods are used in such cases. The seed can be collected by managing the crowns of trees and use of some type of sheeting can be made easily.

(e) **Animal catches:** The animals gather fruits/cones as a source of food supply, but the source is very limited for some areas only. The squirrels are the seed gather, but Sutherland (1979) advocated the danger of infestation by pathogenic fungi that may reduce the germination to a great extent.

The seed collection operation does not take care of the source of seed. This happens when it is done in haste. This may lead to the production of low quality trees. Khan (1987) opined that the seed should be collected from the trees with straight boles, less branches, good growth, free from the diseases, or the pests, healthy and vigorous trees. Therefore, plus trees should be identified and used for a source of good seed for a sound progeny. Sagwal (1978) advocated the same idea while working on the plus trees of *Pinus roxburghii* sergeant in Himachal Pradesh (India).

2. Collection from the crowns of felled trees

This is a very good method of seed collection. This is carried out in the commercial felling during the seed ripening season and to collect the fruits/seeds from the felled trees (Morandini 1962). Further, Douglass (1969) suggested that the seed collection should be carried out when felling in the area is complete. In this method,

(a) Selection;

(b) Marking;

(c) Fellings; and

(d) Harvesting the fruits/cones/seeds must be looked into with care well in advance.

3. Seed collection from standing trees with access from the ground

(a) **By hand:** In case of small trees (Shrubs) or low-branches trees, the seeds/fruits can be picked out directly from the branches. The seed collector can collect the seeds by standing on the ground itself. However, Stein et al. (1974) suggested that the smaller fruits/seeds are generally harvested directly into a basket, bag, bucket, etc.

(b) **Cutting/breaking/sawing:** When the branches bearing the fruits/seeds/cones, etc., are out of reach of man by standing on the ground, a pole and hook may be used for pulling the branches down within one's reach. Therefore, long-handled rakes, saws, chisels, hooks or pruning shears can be used to pull the fruits/seed/cone-bearing branch lets. In this case, high rigid bamboo, aluminum, or plastic poles (generally 4 to 6m in length) are used for polling down the branches of trees whose seed is to be collected. The method in question, however, is not applicable to the trees with acutely angled branches such as *Eucalyptus tereticornis*.

(c) **Use of rifle:** This method is not used in Indian Forestry but it is prevalent in some of the advanced countries. This helps in severing seed bearing branches.

4. Collection from standing trees with access by climbing

There is a limit to the height to which long-handled tools can be used for collecting seeds/fruits/cones, from the ground. Hence, climbing seems to be the better and only option where trees cannot be felled. Therefore, for smooth functioning, the operation may be described as follows:

(a) Climbing into the crown by way of the bole:

- Climbing with minimum equipment.
- Climbing irons or spurs.
- Use of ladder.

(b) Climbing into the crown directly:

- Ladders: They may be of wood, aluminum, etc.
- Ropes and hoisting equipments.
- Climbing and picking fruits/seeds/cones within the crown.

For climbing purposes, it is essential to know the use of equipments for climbing safely, these are discussed below:

- Safety belt and strap;
- Safety line; and
- Safety ropes.

5. Collection from standing trees with other means of access

The system may prove most profitable in a forest stand where repeated collections and made. Following are practiced for seed collection under this method:

(a) Use of balloons-not practical.

(b) Use of helicopter-a good preposition, but too expensive. Not feasible under India conditions. The reasons are quite obvious and beyond the purview of the present reference.

Seeds Storage

Between seed collection and seed storage, there is one very important operation which is known as 'seed processing'. This comprises of the 'seed extraction' process. The process by which the seed is just separated from the fruit is called 'seed extraction'.

However, the purpose of extraction and associated processes is the maximum production of clean seed having high viability (Stein et al. 1974). The information will be given in a very brief account about the seed processing/seed extraction and the process involved here usually include one or more of the following as discussed below:

1. **Maceration:** The seeds must be stored temporarily and may also be spread on some tarpaulins or other sheeting. The following operations must be carried/ performed:
   (a) **Pre-cleaning:** The cones and fruits and the seeds are required to be cleaned of twigs, bark, foliage, and other impurities before they go for extraction, cleaning, storage, or sowing. The impurities take up needless space, and, therefore, in small operations, the major debris may be removed by hand (Turnbull 1975C).
   (b) **Pre-curing:** The pre-cleaning refers to, the deliberate storage and slow air-drying of fruits and contained seeds in order to render them more suitable for subsequent operations of kiln drying, extraction and long-term seed storage. However, the processes assisted by pre-curing are seed maturation and fruit drying.
2. **Methods of extraction:**
   (a) **Depulping:** The depulping of flashy fruits should be time soon after the collection of seed to avoid fermentation and heating.
   (b) **Drying:** The drying using either natural or artificial heat source. It is carried out as under:
   - **Air-drying:** Most dump seeds, cones, or pods/fruits can be air-dried. They are placed on protected drying screens/or trays that allow sunlight and air to pass through, but keep out rain. The drying screens and racks are the basic seed processing equipment. However, the seeds should never be left in the trays/racks for longer period. The seeds are exposed to full-sunlight for only 2 or 3 days.

     For a uniform drying, the seeds are evenly spread over the screens. The seeds must be covered during the evening hours to avoid dewfall accumulation. All seeds must be screened from rodents and birds during drying.
   - **Kiln drying:** Large quantities of seeds, cones or pods are usually kiln-dried. The kilns have many sizes and shapes. Two general types are;
     (*i*) The rotating-drum kiln; and
     (*ii*) Large conveyor-belt progressive-type kiln.
     (*i*) The drum kilns are fairly efficient because their rotating/or tumbling action also separates the seeds from the cones/or pods. The seeds are removed from drum at the earliest as the excessive tumbling is harmful.
     (*ii*) The large kilns are heated by natural gas or electricity. However, when kiln-drying cannot reduce seed moisture content satisfactorily, then the seeds are dried further in the simple ovens. However, the kiln temperatures should not exceed 54°C for pine tree species.

- **Solar dryers:** These are used in those areas where the number of day with full sunlight is high under normal sunlight conditions, the pods and cones usually open with 3 to 4 days. If, the cones are improperly stored while waiting for drying, they may mold. The extraction may be done with a power-driven tumbler inside a solar dryer. The solar dryers are very simple to build, a wide variety of designs exists (Minko and Craig 1976; Bin et al. 1980).

  Turmbull (1975C) classified kilns under the following groups/types:

  (*i*) Stationary tray kiln;

  (*ii*) Vertical progressive kiln;

  (*iii*) Horizontal progressive kiln;

  (*iv*) Rotating drum kiln; and

  (*v*) Portable kiln

- **Separation:** When the cones open after obeying, some seeds full out easily as a result of manual stirring, rotation in rotating drum kilns, or, in certain vertical progressive kilns, from the shaking of the cones as they fall from one tray to another.

- **Vigorous treatments:** The cones can be shaken in course sieves to release the seeds, but more vigorous treatments are required for some tree species.

  (*i*) Trumbling: The treatment is used for conifers (softwoods).

  (*ii*) Threshing: The treatment is used for broad leaved (hardwoods) trees species.

Operations after Extraction

After seed extraction, several operations are required before they are fit to go into storage operation. Following are considered in this operation:

1. **Dewinging:** This is necessary in a number of tree species in order to make seed processing and nursery sowing easier, the wings are usually removed whenever, they are larger than the seed.
2. **Cleaning:** The main features by which sound seeds may be distinguished from inert matter (including sterile and empty seeds). Turn bull (1976C) has given an account of cleaning grading methods which is dependent upon:

   (a) Screening/sieving;

   (b) Sorting according to length;

   (c) Blowing;

   (d) Liquid flotation.

   There are two basic methods used for liquid flotation:

   - Density method; and
   - Absorption method.

(e) Friction cleaning;

(f) Other cleaning methods, several other methods of seed cleaning have been used on trial basis, but are not yet in wide spread operational use. They have been described by Klein et al. (1961) and omen (1969). The methods are as follows:

- Electronic and electrostatic separations;
- Magnetic separators;
- Electronic colour separators; etc.

Grading of Seeds

The grading of seed is very important in the uniform nursery stock. The methods of grading seed vary little from those used in the cleaning process. The following methods can be used easily and effectively for size grading of tree seeds:

(*i*) Sieving and screening;

(*ii*) Cylinders;

(*iii*) Air blowing;

(*iv*) Flotation; and

(*v*) Specific gravity separation.

The seed storage is defined as, the preservation of viable seeds from the time of collection until they are required for sowing (Holmes and Bnszewicz 1958). However, when the seed for afforestation is, supposed to be sown soon after seed collection, no charge of seed is required.

Storage Period

The storage of seeds can be done for varying periods such as;

1. **Upto one year:** The seeds are stored for one year when both the seed production and afforestation are the regular annual features/events, but it is necessary to wait for the best sowing season.
2. **1–5 years or more:** When a species bears an abundant seed crop at intervals of several years and enough seed should, therefore, be collected, in a good seed year to cover annual afforestation requirements in intermediate years of poor seed production.
3. **Long-term storage:** The seeds are stored for the purpose of conserving genetic resources. The period of storage, however, may vary according to the longevity of the seed and the condition of storage.

Longevity of Seeds

1. Evart (1908) divided the seeds into three biological classes:

(a) Micro biotic: The life of seed is upto 3 years.

(b) **Mesobiotic:** The life of seed varies from 3 to 15 years.

(c) **Macrobiotic:** The life of seed is from 15 to more than 100 years.

2. However, Roberts (1973) recognized two major classes of seeds which are as follows:

(a) **Orthodox:** The seeds which can be dried down to a low moisture content of about 5% (wet basis) and can usually be stored for longer periods.

(b) **Recalcitrant:** The seeds which usually survive drying below a relatively high moisture content (20-50%, wet basis) and which cannot be stored for longer period of time.

Factors Affecting Longevity

There are many factors which affect the longevity of seeds in storage, viz.,

1. Seed conditions:
   (a) Seed maturity;
   (b) Parental and annual effects;
   (c) Freedom from mechanical damage;
   (d) Freedom from physiological deterioration;
   (e) Freedom from fungi and insect-pests; and
   (f) Initial viability.
2. Storage conditions;
3. Aging of seeds;
4. Storage atmosphere;
5. Seed moisture content;
6. Storage temperature;
7. Light.

Choice of Storage Methods

Following few points are considered as beklow.

1. Storage at ambient temperature and humidity;
2. Dry storage with control of moisture content but not of temperature;
3. Dry storage with control of both moisture content and temperature;
4. Dry storage for long-term conservation;
5. Moist cold storage with control of temperature;
6. Other storage methods.

Some other methods of storage were used frequently in the past, but are not yet of the commercial application. However, they comprise of the following (Magini 1962; Stein et al. 1974):

(a) Seed storage in running water;

(b) Storage of seed under partial vacuum;

(c) Coating individual large seeds with paraffin or latex to prevent moisture exchange;

(d) Storage of seeds in gases other than air, for example, nitrogen (N) or carbon dioxide (CO_2).

Chapter 5

SEED TESTING

INTRODUCTION

The tests of the quality of seed need to be performed at many stages from parent tree to the nursery bed. Tanaka (1984) advocated that several tests of seeds assess the physical and biological aspects of seeds. However, there are reports that many tests are now standardized at international levels refusing better communication among the nurserymen, foresters, and the scientists who are involved in the study and handling of seeds. The seed tests are generally performed immediately following the seed extraction and shortly just before the actual sowing is done in the nursery.

Justice (1972) head defined the following objectives for developing rules for seed testing such as:

1. To provide methods by which the quality of seed samples can be determined accurately;
2. To prescribe the methods by which seed analysts working laboratories to different countries throughout the world can get uniform information and results;
3. To relate the laboratory results in so far as possible, to the planting values;
4. To complete the tests within the shortest period of time possible, commensurate with the above referred objectives; and
5. To perform the tests in the most economical manners.

Turnbull (1975d) put his ideas that the essence of good seed testing is the application of reliable standard methods. The probable seeds tests are purity, seed weight, germination, moisture content and seed health.

SAMPLING

In seed testing operations, the sample of seed should be the representative of the whole. Bonner (1974) opined that completely homogeneous seed lost would be easy to sample, but they do not exist. It is a famous notion that a single seed log may be small or large, and it can fill the whole cold storage room or one small bag. Therefore, all tests must be carried out by representative samples of the whole seed lot. The seed samples should be drawn very carefully, otherwise the time spent for testing the seeds carelessly drawn samples may be time wasted (Carter 1961).

MIXING

In care of small seed lot, if is very much possible for improving its homogeneity by thorough mixing up the whole lot, but in case of large seed lots, it seems to be quite impossible. It may be due to the following reasons:

1. Large number of labour required.
2. May not become homogeneous due to bulk volume of seed; and
3. It is a time consuming exercise.

Paul (1972) reported that the methods of mixing the seeds are equally applicable to the small seeds lots and that too to the composite seed lots.

Methods of Mixing

There are two simple methods to mix the seeds thoroughly, viz.,

1. **Use of mechanical divider:** These are quite useful in reducing the size of seed lots by repetitive having. The seeds can be mixed as follows:
   (a) Pass the whole seed lot through divider;
   (b) Take two equal portions and put them side by side into the divider;
   (c) The above operation (number ii) may be repeated twice to get good results;
   (d) There after by taking two equal portions, put them into the storage container.
2. **Hand mixing:** It is a simple and erode method of mixing the seeds. The whole seed lot is spread on the cloth or plastic sheet and then mix by scooping the seed from the side and from top to bottom.

Again spread the seed evenly at the same place and divide the whole lot into four equal parts. Put them in the separate four containers of good quality. Thereafter, the seeds are put into the storage container. According to USDA (1974) reported that when seeds are stored in the small bags, mix them carefully before sample. However, for small lots of 1 to 6 bags, sample each; for more than 6 bags, the rule of thumb is to sample 10% of the total number of bags plus 5, as shown below:

No. of bags in seed lot	No. of bags to be sampled
8	6
10	6
20	7
35	9
50	10
100	15

For a lot size of 8 bags, 10% of 8 comes out to be 0.8 or say 1.0, if, it is rounded off, then 1 + 5 = 6 bags to be sampled.

Similarly, 10 bags,

$$\text{Calculation} = \frac{8}{100} \times \frac{10}{1} = 0.8$$

$$= \frac{10}{100} \times \frac{10}{1} = \frac{100}{100} = 1.0$$

Therefore, $1 + 5 = 6$

Weight

The measurement of seed weight is made weight is made on the pure seed component separated by the purity test. The weight is usually expressed as the weight of 1000 pure seeds. The weight may be determined simply by counting 1000 seeds and then weighing them (Bonner 1974, Paul 1972).

The factors affecting the seed weight are:

1. Size;
2. Moisture content; and
3. Proportion of full seeds in the lot.

The counting of seeds for weight can be performed by hand in general practice. Magini (1962) gave the description of seed counting devices which is as follows:

1. **Counting boards:** These are very useful for large seeds, but of regular shape, e.g., Pinus of some leguminous species. The counter consists of two boards, viz.,

   (a) A stationary top board with perforations; and

   (b) A thin solid board below which service as a false bottom.

   A quantity of seed is placed at the corner, the board tilted and moved so that one seed may go into each perforation. The board is then emptier and the known number of seeds from it weighed. However, in other kinds, the counting apparatus is perforated by holes very small to allow the seeds to pass through, but suitable for allowing the flow of air to suck the seeds into position (Tiren 1948)

2. **Vacuum counters:** These are used with the seed of various sizes and shapes. Various difficulties may occur for very small and irregularly shaped seeds. however, these are not meant for very large seeds, e.g., many nuts (Juglansregia)

   The vacuum counting device has three main parts;

   (a) A vacuum system including pipes;

   (b) Counting heads; and

   (c) A quick release valve.

3. **Electronic counters:** In the electric seed counters, a vibrating feed bowl moves the seeds past an electric eye in a single file. The machine can be programmed

to count some sample size, then cut itself-off (Bonner 1974). Manual counting boards, combined with careful random sampling, are likely to be more accurate, although considerably slower.

The 1000 pure seed weight can be converted to seeds per gramme or per kilogramme as follows:

$$\text{Number of seeds per g} = \frac{1000}{\text{Weight in g of 1000 seeds}}$$

or

$$\text{Number of seeds per kg} = \frac{1000 \times 1000}{\text{Weight in g of 1000 seeds}}$$

If, the sample counted is other than 1000 seeds, the appropriate formula is:

$$\text{Number of seeds per g} = \frac{\text{No. of seeds in sample}}{\text{Weight of sample in g}}$$

or

$$\text{Number of seeds per kg} = \frac{\text{No. of seeds in sample} \times 1000}{\text{Weight of sample in g}}$$

To convert the weight of 1000 seeds in grams to number of seeds per pound:

$$= \frac{453{,}600}{\text{Weight of 1,000 seeds in grame}} = \text{Number of seeds per pound}$$

Similarly,

$$\text{Number of seeds per gramme} = \frac{\text{Reported 1000 seeds weight in g}}{1000}$$

Purity Analysis

The purity tests determine the percentage of true seeds and some other material in a seed lot. The tree seed samples may contain following types of impurities:

1. Weed seeds;
2. Seeds of other tree species;
3. Detached seed structures;
4. Leaf particles; and
5. Other material.

The main objective of purity analysis is, to determine the composition by way of weight of seed sample being tested. When purity analysis is done, it is supposed to be the first test to be undertaken as the subsequent test are made only in pure seed component.

What is Pure Seed?

Pure seed refers to that component of a seed lot which consists of seeds of the designated species. However, according to ISTA rules, it includes not only mature, undamaged seeds, but also under-sized, shriveled, immature and germinated seeds provided they can be positively identified as the designated species, and the pieces of seed resulting from breakage which are more than half their original size. Excludes the seeds of other species, wings of coniferous seeds, seeds of coniferous or leguminous species with seed coats entirely removed, broken seeds particles less than half the original size and other matter such as stones, twigs.

ISTA rule (1976) prescribed that weighing be tone in g (gramme) to the minimum number of decimal places which are necessary to calculate the percentage of its component parts to only one decimal place.

The percentage of pure seed is just calculated as follows:

$$\text{Purity \%} = \frac{\text{Weight of pure seed}}{\text{Total weight of orginal sample}} \times 100$$

However, each component is generally expressed as % age by weight of the original sample. It can be illustrated by supposing. The initial weight of Pinus spp., seed sample was 60.124g and the pure portion weighed 52.467g, then purity of the seed lot will be,

$$\text{Purity \%} = \frac{52.467}{60.124} \times 100 = 87.3\%$$

(The example was taken from the book by Leon H. Liegel and Charles R. Venator on Forest Nursery Management).

However, the pure seed component from the purity analysis can be further sub-sampled for the germination test, and also for determining the seed weight. Turnbull (1975 d) pointed out that the production potential of a seed lot can be determined when the following are considered together:

1. Purity analysis; and
2. Germination tests.

Moisture Content

The seed moisture content has an inverse relationship with seed longevity and germination capacity. It has been experienced that at high seed moisture (more than 18% moisture), the seed germination and longevity are at lower level. A periodic checking of seed moisture is carried at regular pace (say every 6 months, if, the seed stored for longer period).

ISTA (1976) developed air oven-method for determining the seed moisture content. It is used for all forest tree species and shrubs except the following ones, viz.,

Fir (*Abie spinolrow*); Deodar (*Cednus deodara*); Beech (Fagus spp.); Spruce (*Picea smithiana*) and Hemlock (Tsuga spp.) these seeds of all these species of trees generally contain oils and resins which are volatile at 105°C. (which is dangerous conditions), Liegel and Venator (1987) described the air-oven technique, a representative seed sample is weighed accurately and dried at 103° ± 2°C for 17 ± 1 hour and this is again weighed after cooling the same. The moisture % is calculated by dividing the weight of water lost in drying process by the wet weight before drying the seed. The difference should not exceed 0.2%. It can be practically illustrated as; suppose that wet weight of duplicate samples A and B are just 20.197 and 20.186g. After 17 hours, dry weight for A and B are 18.062g and 18.002g respectively.

The seed moisture contents are calculated as under:

$$\text{Sample} \qquad A = \frac{20.197 - 18.062}{20.197} \times 100 = 10.6\%$$

Similarly,

$$\text{Sample,} \qquad B = \frac{20.186 - 18.002}{20.186} \times 100 = 10.8\%$$

The difference of moisture of sample A and B comes out to be (10.8 – 10.6) = 0.2%. Therefore, there is no need of re-test the seed samples. On the other hand, if the seed moisture contents are more than 17%, then,

Then seed samples are weighed, pre-dried at about 55°C for only 5 or 10 minutes, it is reweighed after cooling and then air-oven technique is applied. Therefore, percentage of moisture is determined by dividing the weight of water lost by the initial wet weight.

For example, suppose, initial wet weight of a seed sample is 26,000g and that an electronic meter showed a seed moisture content of 32%. To check this (Liegel and Venator 1987), the seed sample is dried at 55°C for about 10 minutes. The resultant weight (W) of 20.197 is an initial loss of 5,803g (W_1) and then after drying for 17 hours or so, at about 103°C, the second weight is 18.062g, an additional loss of 2.135g (W_2). Therefore, actual seed moisture content of the sample may be as follows:

$$\text{Moisture content} = \frac{\text{Weight loss } (W_1 + W_2)}{\text{Initial wet weight}} \times 100$$

$$= \frac{5.800 + 2.135}{26000} \times 100 = 30.5\%$$

Therefore, in the present context, the oven-drying at 103°C for 17 ± 1 hour seems to be satisfactory for calculating the seed moisture content of most of the species. The accuracy at every step is most urgent to get promising results.

GERMINATION POTENTIAL

It is determined from the pure seed component of purity tests. Howcroft (1973) reported that alcohol is used to separate viable from the non-viable seeds. Also, AOSA (1981), according to this association, when the seeds are not available then representative samples of lots must be tested. This estimated should be at least 98%. However, Bonner (1974) reported that none is more important that the potential germination of the seeds.

Definition of Germination

1. Germination is defined as the emergence and development from the seed embryo of these essential structures which are indicative of the seed capacity to produce a normal plant under favorable conditions (Justice 1975; ISTA 1974).
2. In general, resumption of active growth in the embryo of a seed as demonstrated by the protrusion of the radicle.

Germination is usually expressed as the percentage of pure seeds which produces normal seedlings or as the number of seeds germinating per unit weight of the sample. Following are the environmental factors which influence the germination of seeds such as:

1. Moisture;
2. Temperature;
3. Aeration; and
4. Light.

Test Procedures

In germination tests, the seeds are thoroughly mixed and the representative samples are drawn by hand at random. At least, 600 seeds in 6 replications are made. The replicates may be small say 50 or 25 seeds may also be used. The mean germination as expressed for all replicates can be used for calculating the number of seeds required for actual sowing of seeds. in some cases, cold stratification or seed coat scarification are also carried out and Gupta and Kumar (1977) reported that some seeds require some temperatures for best germination results.

Some germination tests are still carried outdoors in floats containing sand, peat, vermiculite or the soil. It has been noticed that outdoor test have added risks of damages/or loss of seeds by

Also soft-X-rays tests are faster than the staining tests and indicate insect/ or processing damage as well. Kamra (1973), the x-ray technique has been used successfully for Teak (*Tectona grandis*).

Other Special Tests

Many other seed germination tests may have implication for some particular nursery operations as discussed below.

1. **Seed vigour:** The vigour is a loose term used to describe observable germination differences in seed lots of similar/or different genetic make-up. The common expression of vigour for tree seeds includes assessing speed of germination, also known as germinative energy. The germinative energy is usually defined, that proportion of germination which has occurred upto the time of peak germination, or the time of maximum germination rat, or upto same pre-selected point, usually 7 test days.

   Also, germinative energy is defined as, the percentage of germination at the peak rate of germination. However, such definitions are applicable for long-term observations or experience of nurserymen, but these are seldom quantified through repeated and replicated trials.

2. **Indirect indices of seed vigour:** In real sense, indirect indices of seed vigour include. Biochemical staining tests by using tetrazolium (TZ). However, many vigour classes may be developed, generally based on partial/or complete staining of embryo and the place/location of damaged or even dead tissues. Perera (1973) reported that a successful technique exists for Eucalyptus spp. (Eucalypts).

Chapter 6

PRE-SOWING SEED TREATMENTS

INTRODUCTION

Seeds, under favorable conditions of environmental factors of moisture and temperature germinate quickly. It has been visualized that some tree species have degree of seed dormancy. If, dormancy is strong enough, some kinds of seed pre-treatment is urgently required. This facilitates high germination rate in a shortest period of time. The example, in Robinia (*Robinia pseudacucia*) the best treatment gave a ten-fold increase in germination after 10 days in comparison to the control. Bonner et al. (1974) and Forest (1967) gave similar types of observations in *Pinustaeda* and *P. elliotii* respectively. Further, the decision whether or not to pre-treat the seed with vary not only with tree species but also with the following:

1. Provenance condition;
2. Seed year;
3. Local nursery conditions; and
4. Length and condition of storage

Pre-treatment is supposed to be very important for,

TYPES OF DORMANCY

Dormancy is defined as, a physiological state in which a seed predisposed to germinate does not, even in the presence of favorable environmental conditions.

Dormancy of seed is of several types and sometimes, more than one type may occur in the same seed. In the simplest forms, the dormancy may be of following types in general:

1. Exogenous/or seed coat/or pericarp dormancy;
2. Endogenous/or embryo dormancy; and
3. Combined dormancy.

Gordon and Bowe (1982), the types of dormancy have been differentiated as follows:

A. Exogenous dormancy	
Physical Chemical Mechanical	impermeability of seed coat or pericarp to water. inhibitors in pericarp or seed coat. mechanical resistance of pericarp or seed coat, to the embryo growth.
B. Endogenous dormancy (Morphological)	
Morphological	under-development of embryo.
C. Endogenous dormancy (Physiological)	
Physiological	i.e. germinated by physiological inhibiting mechanisms.
Shallow	Inhibiting mechanism weak.
Intermediate	Inhibiting mechanisms intermediate
Deep.	Inhibiting mechanism strong.
Combined morpho-physiological dormancy	
Combination of underdevelopment of embryo with strong physiological inhibiting mechanism.	
Combination of underdevelopment of embryo with strong physiological inhibiting mechanism of epicotyl growth	
Combined exogenous/Endogenous dormancy	
Various combinations of seed coat or pericarp dormancy with endogenous physiological dormancy, e.g., Tiliacordata combines physical impermeability of the seed coat with deep physiological dormancy.	

1. In *moist tropical forests*, dormancy is not a problem. The conditions of temperature, moisture and oxygen are suitable for germination just after on the occurrence of seed dispersal. Therefore, many a species germinate with a shot span of time, say within days or weeks and will not benefit from dormancy.
2. In *dry tropics*, the seed coat dormancy is quite common and pre-treatment is very urgent to break the dormancy to facilite the rapid and uniform type of germination of seeds. Both physical dormancy which is usually caused by hard seed coats/or pericarps with layers cutinized which are supposed to be impermeable to water. Similarly, chemical dormancy which is caused by inhibiting the chemicals present in the seed covering.
3. In *temperate forestry*, methods of overcoming the various types of endogenous dormancy have been investigated for a long period of time. However, the physical and chemical types of exogenous dormancy are likely to be of most concerns for everyone whose are engaged in this profession.

Breaking Seed Coat Dormancy

The dormancy of seed coat is just corrected by breaking the hard impermeable coat which usually surrounds the seeds, for example, Teak (*Tectona grandis*) and leguminous plant/trees belonging toliguninosae family by mechanical/or chemical methods.

Acid Scarification

Sheikh (1979) reported that a common chemical method is scarification with commercial grade sulphuric acid (H_2SO_4). Various safety precautions are used while dealing with acids soak methods such as:

1. Using acid resistant wire containers or the screens for handling;
2. Draining and washing seeds;
3. Providing a good seed source of running water;
4. Establishing a safe area to drain and dilute the acid obtained from rinsing; and
5. Finding an area to dry seeds after rinsing.

It has been observed that some species require 15 to 60 minutes of soaking whereas other species require upto 6 hours. The acid-soak treatment that yields the highest percentage of swollen seeds, (many be caused by water uptake) is supposed to be the best treatment.

1. Lots upto 100 kg are treated effectively by placing a wire screen filled with seeds inside a metal drum.
2. Another alternative for thin-coated seeds are placed them in conical pile on a flat, hard surface, apply acid at 2.5kg per 36kg of seeds. These are mixed vigorously thoroughly mainly with a shovel, rinse and the treated seeds.

Advantages

Following are the main advantages of acid scarification:

1. It requires little or no special equipment's in this method;
2. It is a cheap method and hence low-cost; and
3. The recovery of acid for future prior-treat.

Disadvantages

It has following disadvantages, viz.,

1. The length/duration of treatment must be carefully determined;
2. Temperature plays/duration of treatment must be controlled, mainly when large seed lots be controlled/protected; and
3. It is a well-known fact that the acids are hazardous substances. Acids may be harmful to human-beings.

Precaution

When the acid is used, never add or splash water on acid because a violent reactions or explosion can occur.

Water soaking: A large number of treatments involve soaking the seeds in water/ or other liquids. (Kemp 1975) reported that some seeds with very low resistance to germination can respond well to soaking for 24 hours in water at an ambient temperature, Mafias et al. (1973) also found that 48 hours soaking of *Pinus caribaea* seeds in water (say at room temperature) resulted in more uniform germination than in the controlled condition (untreated seeds).

However, in India conditions, soaking of seeds in water (simple water and not the treated one), for varying time periods from 2 to 48 hours. This enhances the rate of germination to a satisfactory condition in the following tree species:

Acacia meamsii, *A. melanoxylon*, *A. nilotica*, *Albiggi zamara*, *A. procera*, *Grevilea robusta*, and *trewia nudiflora* (reported by pattamath 1982) .

Teak (*Tectona grandis*) is the versatile tropical tree species where there is an evidence that chemical dormancy is more significant than the physical dormancy. Fairlamb and Davidson (1976) found very interesting fact that an aqueous extract, got by soaking the seeds/fruits of Teak (*Tectona grandis*) for about 4 days inhibited germination of cress seeds, following are observations:

Germination was, 11 in 144 hours soaking in extract; 76% in tank water; and 96% in distilled water.

However, it was found interesting that there was no evidence for physical dormancy in Teak (*Tectona grandis*). But it is evidenced that there was a great variation in dormancy degree between the provenances of teak (*Tectona grandis*).

The leguminous seeds (some only) gave promising results in hot water treatment. Here, the seeds of a particular leguminous tree species one placed/put into the boiling water. These seeds are removed immediately and kept for cooling. The seeds in such conditions may get imbibed and swell as the water coals down slowly. A proper relationship of volume of water to the volume of seeds can be determined by the help of experiment. It may however, vary considerably with respect to the species,

- Ratios of 2-3 times (Goor and Barney 1976);
- Ratios of 4-5 times (Bonner et tal. 1974); and
- Ratio of 5-10 times (Seeber and Hgpaoa 1976).

as much water as the seed have been suggested. Therefore, hot water treatment is relatively easy and very safe to apply, as well as being very effective with some species of trees, Heit (1976 b) reported that it is no well adapted to larger lots because of the difficulty in handling and sowing the swollen seeds.

Biological Methods

The animals and micro-organisms are very crucial and play on important factors in breaking-down the seed coat impermeability. However, it seems to be difficult to use these organisms as a controlled pre-treatment of seeds, but exceptions are always there, and in a few cases very successful results were obtained.

The seeds of *Acacia sengeal* and *coratonia siliqua* have been passed through the digestive tracts of goats, germinate quickly and with good percentage. The seeds of some tree species are said to be regurgitated after partial rumination, e.g., *Gmelina arborea* (Greaves 1981). Stated that seeds of *Acacia nilotica* (babool/kikar) are ejected after rumination by sheep and goats but pass right through the digestive tract in the cattle.

The termites are an important agent for breaking the seed coat dormancy in the tropics. These have been used to break-down the tough and bristly fruits of *Plerocarpus angolensis* (Groome et al. 1957).

Partial fermentation

It is damaging to many seeds, can be beneficial in over coming the seed coat dormancy, e.g., fruits of Teak (*Tectona grandis*) are spread on the ground through rainy season for partial fermentation.

These seeds are collected and stratified in pit with the layers of,

1. Seeds;
2. Organic matter; and
3. Soil.

These are watered daily for about 10 days for satisfactory germination results (Wunder 1966).

Mechanical Scarification

It has been suggested that impermeable seed coats can be broken mechanically with the files, sand paper, and electric needles for small seed los and with hand or sometimes the motor-driven scarifies, such as sand paper-lined drums and cement mixers for large seed lots.

The germination tests of seeds are required, to determine optimum treating time. However, over-treatment of the seeds can cause damage by resulting in reducing of germination and this increase the risk of fungal attack. Therefore, such eventualities must be avoided at all costs.

Advantages

The advantages of mechanical scarification are as follows:

1. There could be less risk of injury to the workers/laborers;
2. There is no need to control temperature under such conditions; and
3. The seeds remain dry throughout pre-treating allowing immediate sowing after wards.

Disadvantages

1. The seeds are easily damaged by over-treatment;
2. The scarified seeds are more easily damaged by pathogens than that are non-scarified seeds;
3. The large lots require some special equipment's, and tools; and
4. The seeds with resins or fleshy pulp cannot be used in sand-papers-lined trumblers.

Stratification Method

Besides seed coat or external dormancy, some tree seeds have internal dormancy. Therefore, such seeds germinate, they should undergo physiological change, coal, most treatments that artificially may cause such changes are called stratification.

Cold stratification

In cold stratification, the seeds are stored in a well-aerated medium such as sphagnum mass and kept at low temperatures of 1°C to 4°C for to 4 months.

Precautions

The following three precautions that must be taken care of;

1. Maintain adequate moisture which can be imbibed by seeds throughout stratification;
2. Keep temperatures low, near freezing, or to reduce microbial activity, reduce sprouting and stop/prevent heat build up from seed respiration; and
3. Supply sufficient aeration for gas and heat exchange, for long-term stratification of two months and longer.

Definition

The stratification can be defined as follows:

1. The practice of burying seeds in moist medium, often in alternate layers to over come dormancy; commonly applied to any technique which keeps seeds in a cold and moist environment.
2. Refers to, the method of placing the seeds in layers alternating with layers of a moisture retaining medium, such as sand, peat or vermiculite, and keeping them at a cool temperature for a period, which is commonly between 20 and 60 days, but varies considerably from species to species.

However, the use of the word stratification as recently been expanded to include all forms of cold, moist treatment whether or not the seeds are placed in the layers (Bonner et al. 1974). Further, low enough temperatures is used to accomplish pre-chilling are felt by over-winter stratification in the outdoor pits, the methods used are describe by Aldhous (19672), viz.,

1. The pit should be located in a coal, shady and well-drained site. The bottom of pit 10.0 cm should be filled with sand or gravel for the purpose of drainage. The pits should never become water-logged;
2. The depth and width of pits should be about 60-80 cm; however, length can be adjusted to the volume of seeds to be stratified;
3. The pits should be lined at the bottom and sides with a frame carrying 6 mm mesh-wire netting as a protection device against the nice;
4. The seeds must be mixed with 4 times its weight of sand. The pit filled with the seed/sand mixture to within 15cm of the surface. On the top of pit, sand should be filled in the pit.

5. The start of stratification must be timed in relation to the anticipated sowing date (time) and the optimum period of stratification expected for the species.

However, the seeds should be sown soon after the removal of seeds from stratification. The stratified seeds of some tree species, e.g., Prunuspadus may undergo a secondary dormancy, if, subjected to;

1. extreme drying; or
2. to temperatures over 20°C.

USDA (1974) gave many techniques of stratification which are generally used frequently, viz.,

1. **First,** it should be decided whether the seed lot size is small enough for pre-treating in small plastic bags or very large for bags and drums. The seeds are conditioned to high moisture contest by water-soaking for overnight at room-temperature.
2. **Secondly,** choosing a moisture-holding medium such as granulated peat mass, is quite oftenly used, because it is a little bit acidic in nature, and like sand, dispels heat very well. However, the medium must be moist enough.
3. **Third,** the seeds are usually mixed with the treating medium. The whole mixture is placed in some canvas bags and preferably kept in drums. A drainage hole must be provided at the bottom of the drums. However, small seed lots are stored alone in plastic bags. Sometimes, the moss is kept on them. The seed should be packed tightly and hence must be in loose conditions. The drums and bags must be quickly closed for preventing the seeds and medium from the drying out, which causes a great loss. Regular inspection of such bags must be carried out, and the seeds must be tuned frequently. The storage of such bags must be kept at 1°C to 4°C temperature. Therefore, the seeds are removed and washed.

Chemical alternatives

Some chemical means have been investigated by the seeds scientists to eliminate the need for cold or warm stratification. USDA (1974) reported that the inorganic ions, organic acids, growth regulators, e.g., Gibberellins have stimulated the germination to a great extent. The germination of dormant eucalyptus spp. was improved by using gibberellic acid (Krugman 1974).

Seed Dressing and Pelleting

In seed dressing/or seed pelleting treatments, the surface of the coated seeds with some inert material, the sticker, to which chemicals of various kinds may be added. Magini (1962) claimed the advantages of pelleting which are as follows:

1. The incorporation of fertilizer material in the pellet furnishes the young germinated seedlings with required nourishment;
2. The plant growth regulators or stimulants can promote the rooting/or hasten the emergence of the seedlings;

3. The fungicides and insecticides are more effective when in direct contact with the seeds;
4. Seeds may be protected against the damages by rodents just by adding unpalatable repellent or poisonous substances;
5. The small seeds may become somewhat larger and heavier and this may improve the scatter pattern in aerial seeding operations.

However, investigations have shown that some fungicides used in seed dressing may have an effect on germination. Moreover, fertilizers, fungicides, or insectides can be applied more easily to the nursery soil than to the seeds. Pelleting is used sometimes to improve the uniformity of seed for accurate sowing in the nursery.

Other types of pre-treatment

Simak (1981) designed one pre-treatment for separating viable full seed from that of non-viable full seed. It involves a form of pre-germination of the sound seed. The method comprises of the following three stages, viz.,

1. The provision of conditions ideal for just initiating the internal seed processes that may lead to germination in the viable seeds;
2. Partial re-drying of the seeds;
3. Separation, for example, by flotation, of the non-viable from the viable seeds.

A method, which is usually known as the IDS (Incubation-Drying-Separation) treatment, requires very close control of the following:

(*i*) Temperature; and
(*ii*) Moisture conditions.

The method has application in the situations, where,

1. A substantial proportion of full, but non-viable seed is expected after storage;
2. Control of temperature and moisture condition is possible; and
3. The viable seed obtained after separation can be easily sown in the nursery with minimum delay.

Chapter 7

SEEDBED PREPARATION AND SOWING

INTRODUCTION

Care taken in the preparation of ground, sowing and early seedbed monitoring may result in,

1. Better drainage;
2. Fewer disease problems; and
3. More and better seedlings.

The methods to produce the nursery stock include container (Holmes and Floyd 1969; Geary et al. 1971; Marrero 1962; Marrero 1965; Venator and Munoz 1975; and Voorhoeve 1968) and bare-roots (Barres 1964; Marrero 1948; Monteith 1969; and Wadsworth and Marrero 1953), methods. The container methods are most accepted all over the world because the seedlings and cuttings, attached with soil and growth medium may survive very well. The transplants are generally germinated in seedbeds of nursery.

CONTAINER SYSTEM

These systems are used in very difficult conditions for reforestation programmes. It thus been noticed that on non-critical areas/sites, the potted seedlings may prove better for getting higher plant survival and even the good growth of seedlings.

Advantages

1. Good and fertile soils are not required in the nursery areas;
2. It takes shooter time in the nursery;
3. The roots are not exposed to the stroke of heat and air during the transport of seedlings to field/plantation sites/areas;
4. Due to the less transport shock, faster growth is expected;
5. There could be chances of transplanting time extended due to,
   (a) Plants/seedlings with moisture; and
   (b) Nutrient holding medium in the seedlings.

Disadvantages

1. The container systems are somewhat costly;
2. They are time consuming;

3. Big size pots may create problems of storage and transport;
4. A regular source of good medium for potting is required;
5. During repacking operation, there are more chances of severe damage to seedlings, roots;
6. Only less of seedling are transported per trip.

Shape and Size of Pots

The growth and development of seedlings are directly influenced by the shape and size of the pots. Theron (1972) advocated that long pots permit longer and better root systems of the seedlings,. Whereas the short pots may produce short root systems. The size also plays a crucial role in the growth medium to fill up the pots. However, large pots do not always produce the vigorous and bumper nursery stock.

In general, the pines grow 25 to 35 cm fall in 6–8 months, but, if, they are fertilized, then this may take only 4 to 5 months to attain the height of seedlings as 25 to 35 cm. The most common containers are used, made up of plastic. However, other types of pots are clay, split bamboos, tin coms, milk cartons, etc.

Preparation of pot media

It has been reported that most suited potting media should be,

1. Very cheap;
2. Readily available; and
3. Must be economic.

Lowe (1967) and Marrero (1961) reported that the alternatives are,

1. Soil alone;
2. Soil and mixtures;
3. Organic based mixtures with many types of proportions of,
   (a) Peat;
   (b) Compost;
   (c) Sugarcane wastes;
   (d) Rice hulls;
   (e) Saw-dusts;
   (f) Synthetic mixtures.

If, nothing else is available, then sandy to sandy loam soil is supposed to be the best. However, heavy clays are not preferred due to poor drainage of water in such cases. The organic based mixtures are always considered good because of the following:

1. They are light in weight;
2. Possess very good textures;
3. Good water holding capacity; and
4. Good nutrient rention.

The ratios of sand, soil and composts vary according to the local materials used under such conditions. It is advisable that all materials should be free from cloys, stones, and other debris. If, available, methyl bromide or some other fumigants are usually used.

How to Fill Pots?

The bamboo or metal funnels are very helpful in putting the potting medium into the containers. The compacted soil does not have the improved moisture retention due to the greater capillary pore spaces. However, extra/excess medium dropped between pots invite the growth of weeds. The sites of pots should be neat and clean in between the lines and rows of the pots in the nursery.

Direct-seeding Method

The germination percentage tests determine the amount of seeds sown in the particular area/site. In a general rule, about 20% more seeds are sown, then the number of seedlings needed for allowing natural mortality and cutting in the nursery areas.

The direct seeding is nothing, but the placing of seeds on potting medium in the pots. For example, large to medium-sized seeds of pines and ggmelina are the best suited for this method. But the small seeds are not suitable for this method until and unless the small-sized seeds are pelletized carefully.

Advantages

1. It prevents additional steps of germinating in trays and the transplanting into the containers; and
2. If, two or three seeds are sown per container, the number of filled containers is very high.

It is a very good practice that after seeds are sown/put onto the container medium, they must be covered with a very light layer of soil vermiculite, or the rice hulls. This will help to protect them against the animals attack and the damage from irrigation water (may be in the form of drops). The direct seeding may be carried out in the following two methods;

1. By manual operations; and
2. By the help of machines.

Solberg (1978) reported that care must be used in culling excess direct-seeded plants from the pots for avoiding the damaging the root-systems of the remaining seedlings.

Transplant Method

The transplant refers to, the cultural practice of moving seedlings from one bed to another to promote additional growth. Also, a seedling after it has been lifted and then replanted one or more times in the nursery.

The following components are taken into consideration in the transplant method:

1. **Germination medium:** The best media for seed floats are supposed to be,
   (a) Vermiculite;
   (b) Washed and sterilized river sand or sand/soil mixtures;
   (c) Rice hulls; and
   (d) Sometime the tissue papers for very small-sized seeds

- **Sand/soil:** A 1:1 volume mixture of washed sand soil is the most commonly used germinating medium. However, the soil and sand may be used individually. The particle size should be ≥ 1.7 mm, as determined by sieve analysis. The drainage and aeration properties are excellent for seedling emergence and good growth. Sterilization should be done to prevent disease and the weed growth.
- **Rice hulls:** The rice-hulls are the cheap, local substitute for sand/soil or vermiculite germination media. These are very good growth promoters of the seedlings. It has been noticed that the unground rice hulls are light, making moving and lifting of germination quite easy. Moreover, usually, the rice hulls are free of most weeds and the pathogens. Here, the sterilization should be used when there is a danger of contamination.
- **Vermiculite:** The vermiculite is the ideal germination medium because of the following:
  (*i*) It is sterile and free from pathogens;
  (*ii*) The particle size may allow the aeration and moisture retention;
  (*iii*) Reaction of the medium is essentially neutral (pH 7.0);
  (*iv* The seedlings roots growth is unrestricted;
  (*v*) The seedlings generally and easily emerge from underneath a covering layer of the proposed medium.
- **Tissue paper:** It is a well-accepted fact that the seeds of eucalypts and Kadam tree species are very small. They donot grow well, if, they are just placed directly on any germinating media. The seeds of these species are first placed on the tissue paper and it improve the germination. The tissue paper colour/ or the type does not matter as longer as it is clean, light and very porous. There should be one precaution that the fertilizers should never be applied in such cases. The application of fertilizer may cause damping-off disease which proves to be very fatal to the growing seedlings. Therefore, if, should be avoided at all costs.

Sowing and Germinating

For seed sowing, the trays, made up of wood or metal, are generally used. Commonly, galvanized metal trays of size 30 by 45 by 8 cm (30 × 45 × 8 cm) are used for this purpose. The trays are filled with 4–6 cm of the top level, the medium is smoothed and gently tamped. Thereafter, the suits are placed on the medium which is prepared for the purpose. Later, on these are gently covered with 6 to 10 mm of vermiculite, rice hulls, or fine sand (the sand should be very well sterilized). It has been observed that the coverings allow better emergence of seedlings that the soil alone. Moreover, these evenings protect the seeds from the animals, Marrero (1947) observed and noticed that sowing the seeds deeper than 6 to 10 mm below the surface may cause prevention of emergence of seedlings. The large-winged seeds are usually de-winged and then placed on the germination medium.

It is a very common practice that the seeded trays are kept under 20 to 50% shads and watered as and when required. However, excess watering should always be prevented to avoid damping-off disease. The position of watering must be taken care at regular pace.

2. **Bare-root system:** The bare-root nursery stock is usually produced by the following two ways, viz.,
   (a) Labour intensive; and
   (b) Mechanized methods.

The most successful grown species by the bare-root methods are as follows: *Pinus spp*. (Pines), *Tectona grandis* (Teak), *Gmeldina arbera*, *Casuarina equisetifolia*, etc.

Advantages

1. The sowing of seeds directly in the ground-level, or raised beds prevents transplanting after germination;
2. These are the best stock for mechanized planting methods;
3. In this method, bundling, handling and packing are quite fairly simple for treatment;
4. There is always less weight to transport from nursery to planting site/area/field, and thus allowing more seedlings per trip; and
5. Early as well as later tending operations are usually quite easier.

Disadvantages

Inspite of many advantages, there are some disadvantages of bare-root system, viz.,

1. The nursery soils require inherently good physical and chemical properties for permanent use;
2. A large quantum of high-quality irrigation water is required under this method;
3. The seedlings require 2 to 3 months more time for development; and
4. The damages are potentially much greater to the roots, if, at all, they are exposed to the air and heat after lifting.

Orientating of Beds

The orientation of beds has been described elsewhere in the text. However, for large bare-roof nurseries, available land and its topography will certainly dictate the bed orientation.

Follows and cultivation

The follows/or the cover crops are usually grown on the idle nursery land/area to,

1. Protect against erosion; and
2. To build up soil organic matter.

The nursery bed areas/sites, that have been in fallow conditions, should be ploughed time and again for 4 to 6 months before sowing the seeds. This process allows the decomposition of weeds and grasses, which are nuisance to the growth of the seedlings, A C/N ratio of 30:1 or sometimes less than this, will help to avoid the adding of supplemental nitrogenous fertilizers. Somewhere in the forestry literature, the typical ratios (based on dry weight of C and N) are,

Alfalfa = 13:1;

Legume-grass hay = 80:1;

Oat-straw = 81:1; and

Saw-dust = 250:1.

There exists a close relationship between organic matter and nitrogen content of the soils of nursery beds.

The ratio of weight of organic carbon to total nitrogen in the soil is called carbon/ nitrogen ratio. This is popularly known as C/N ratio. In case of soils, C/N ratio is fairly constant. This is very important in regulating the,

1. Available nitrogen;
2. Total organic matter; and
3. The rate or organic decay.

In aerable soils, the C/N ratio ranges from 8:1 to 15:1. The medium being between 10:1 and 12:1, the variation in the C/N ratio is mainly due to climatic conditions. A wide C/N ratio (40:1) indicates little or no decomposition. Therefore, the C/N ratio helps to conserve both the nitrogen and organic matter in the soil. Further, there exists a close relationship between organic matter and nitrogen content (OM/N ratio) of the soils. In a very crude way, OM:N ratio of 20:1 can be ideal for consideration.

It is customary that, before cultivation, the bed areas must be fumigation. This practice is usually carried out in large areas of nurseries and not the small area of nurseries, for big areas, tractors are used whereas for small areas, hand cultivation is practiced which is very convenient method of cultivation.

However, hand cultivation is confined to the beds of 5.0m or shorter lengths. However, the standard beds are 1.2m in width with about 7 (Seven) seedling rows. Here, the beauty of these beds is that the soil can easily be mounded with hand tools. The mounting has several advantages, mainly increased aeration and drainage. Further, it has the advantage of reduced effects of compaction in the bed itself. The walking path should be at least 45 cm wide between nursery beds for easy access for the operation of weeding and protection.

Sowing and mulching

The successful sowing depends upon the following factors such as:

1. Seeders;
2. Sowing date;
3. Sowing depth;
4. Seedbed density; and
5. Sowing formula:
   (a) Seed factors;
   (b) Nursery factors; and
   (c) Calculating sowing formula

In western countries, two most common types of seeders are generally used, viz,

1. Oyjord seeder; and
2. Wind river seed drill.

However, according to the survey, neither is wholly satisfactory-the largest complaint is clumpy seed distribution, a problem which has not been solved by any of the seeders. A comparison of these seeders is as follows (adopted from Lott and Lowman, 1976).

	Character	Oyjord	Stan Hay	Wind River
1.	Ease of adjustment	Excellent	Poor	Fair
2.	Range of adjustment	Adequate	Adequate	Adequate
3.	Ease of calibration	Excellent	Poor	Fair
4.	Clean out	Excellent	Poor	Fair
5.	Seed damage	Low	Moderate	Low
6.	Range of travel speeds	Adequate	Limited	Adequate
7.	Variation with speed	Low	High	Low
8.	Number of hopper	Me	One/row	One with one pocket/row
9.	Depth control	Good	Good	Good
10.	Seed covering	Good	Good	Good
11.	Construction	Good	Good	Good

From the above referred table, Oyjord was found slightly superior.

2. **Sowing data:** The time of sowing generally depends upon the following factors, viz.,
   (a) Timing of seed ripening;
   (b) The rate of growth of species; and
   (c) The size of the plant to be planted out.

The spring sowing may give plants which are supposed to be fit for planting in the field during rainy season. In case where the tall plants are supposed to be planted/or where the stumps are to be made for planting, the sowing of seed is done during rainy season. These plants get proper size to be planted in next rainy season. However, there is a danger that the plants become very tall (robust type) in the period of one year, the rate of growth may be checked by the following factors, viz.,

(a) By late sowing;

(b) By using less watering;

(c) By using less quantity of leaf mould, F.Y.M. (Farm Yard Manure), or some chemical fertilizers;

(d) Transplanting/or pruning of roots/or shoots.

3. **Sowing depth:** Sowing depth is a crucial factor to produce a uniform bed of seedlings. The selected depth of seed sowing which can be achieved, may usually depend upon the following factors:

   (a) Uniformity of the prepared seed beds; and

   (b) Soil properties.

   It has been observed that, if, a bed lacks level, then sowing depth will vary greatly from row to row. However, this variation may result in the following

   (a) Lessen germination;

   (b) Referred growth; and

   (c) Reduce crop uniformity.

   The best germination is obtained by sowing seed only as deep as necessary to cover it. This may prevent the erosion or the birds from removing it. The most commonly accepted depths range from 0.31 to 0.62 cm.

4. **Seedbed density:** The density at which a nursery chooses to sow may depend on the seedling characteristics desired at harvest. The seedling spacing must be as uniform as possible for getting optimum growth. However, at the lower densities commonly used are, 108 to 2015 seedlings/m^2 or say 10 to 20 seedlings/ ft^2, A density is usually chosen which may give maximum number of seedlings per square meter at the minimum lost. Somewhere else, a bed density is selected based on seedling specification, for example, 4.0 mm diameter, on an acceptable compromise between maximum yield percent (%).

5. **Sowing formula**

   (a) Seed factors: It is very urgent and essential to know the quality of the seed lot which is supposed to be undertaken for sowing. However, the seed lot quality is determined on the basis of three main factors, viz.,

   - Seed purity per cent (%) is the percentage of seed lot by weight that is seed not debris. Therefore, a high percentage (25%) of debris can cause seeders to reduce seedling accuracy.

- **Seed germination per cent (%) is the percentage of the total number of seeds tested which usually germinate after a standard treatment and a set period of time.**
- **The pure live seed, the expected number of germination to be produced per kilogram can then be calculated. The difference between laboratory and field germination, if, known, can be used to modify the pure live seed calculation.**

(b) Nursery factors: The nursery factors which affect the sowing formula include;

- Tree yield;
- Shippable percent(s);
- Bed density; and
- Number of seedlings ordered.

The tree per cent depends heavily on cultural practices and environmental influences.

(c) Calculating sowing formula: Here, once the seeds and nursery factors have been determined, the calculation of sowing formula becomes somewhat easy.

- Pure live seed per kilogram (PLS kg)

$$= \left(\frac{\text{Germination \%}}{100}\right) \times \left(\frac{\text{Purity \%}}{100}\right) \times \left(\frac{1000\,\text{g}}{1000\,\text{seed cot}}\right)$$

- Number of distributable seedlings produced per kilogram of seed

$$= \frac{\text{PLS}}{\text{kg}} \times \left(\frac{\text{tree \%}}{100}\right) \times \left(\frac{\text{yield \%}}{1000}\right) \times \left(\frac{\text{shippable \%}}{100}\right)$$

- Kilograms of seed required = (No. of seedlings ordered) / (No. of seedlings/kg).
- Nursery space required (meters) = [(Kilograms of seed required) × (PLS/kg) × (tree %/100)]/ Bed density.

(Adopted from B.E. Thomson; In: Mary L. Duryea and Thomas D. Landis: Forest Nursery Manual).

Mulching: After seeding, the mulches are generally over the seedbeds to protect the seeds from.

1. Animals;
2. Erosion hazards;
3. Weed infestation; and
4. Conserve soil moisture.

Solberg (1978) reported that mulching may or may not influence the seedling survival in the nursery through outplanting time. Mulching is a practice of covering the soil surface with organic materials such as show, grass, stones, plastics, etc. This reduces evaporation, to keep down the weeds, and also to moderate diurnal soil temperatures. The soil and run-off losses can also be reduced to a great extent to the soil surface.

Though, the mulches are very useful in mitigating the moisture stress effects, availability and the cost is limiting their use in the nursery beds. Reddy (2004) reported that the surface water infiltrate into the soil, the colloidal soil particles are just filtered out in to surface layer. This process, however, simplify the flow of water into as well as through soil pores which lead to surface scaling and soil compaction.

A modification of traditional mulching called Vertical Mulching has been developed for heavy type of soils where there is infiltration problem. In such cases, tranches are dug at 5-10m intervals mainly depending upon the slope at sizes of 30 × 60 cm just across the slope on grade. The, however, are filled with stalk materials that keep the cracks open and allow better water intake which is very much beneficial to the growing seedlings in the nursery beds.

Chapter 8

TENDING OF SEEDBEDS AND CONTAINERS

EARLY TENDING

The container and the bare-root system have almost some only seedling needs/ requirements for water, shade and weeding. The tending of early stage may include. The first four to six weeks of growth of seedlings (germination). It is more expected in such cases the sufficient protection of seedlings, post-control, and the nutritional requirements are generally monitored.

1. Container System

(a) **Transplanting:** These are taken up as under:

The transplanting is usually carried out after seedlings become large enough. But they be handled easily and safely and should not be crowded in the trays. Venator (1973) advocated that the optimum time to transplant the germination of pine tree species is 3 to 6 days.

The lifting of seedlings at an easily stage serves mainly two following purposes, viz.,

- The seedlings should be lifted before the commencement of top roots that usually develop lateral branches; and
- It has been noticed that the large roots bend very easily during the course of transplanting which may develop J-type roots which is fatal to the seedlings.

Further, transplanting is Deon into the lined-out pots exhibit two prominent advantages in comparison to direct-seeding, viz.,

- A satisfactory density of seedlings can be achieved, if, only one seedling is planted in the single pot; and
- When the seedlings are just ready lifting the transplanting may be staggered for many days or even weeks.

There is always a risk in lifting several hindered seedlings from germination trays and placing them in a common plot of water for avoiding root desiccation is a direct and open invitation to damping-off disease (a fatal disease) and some other fungal diseases of seedlings. If, dumping-off disease is noticed, the treatments and precautions must be followed as under.

Treatments to prevent dumping-off

	Type of treatment (1)	Seeds (2)	Germination medium (3)
A.	Steam sterilization (before seeding)	×	✓
B	Fumigation/before seeding		
	1. Methyl bromide gas	×	✓
	2. Phosphine	✓	✓
	3. Ethylene dichloride	×	✓
	4. Carbon tetrachloride	✓	×
	5. Carbon disulphide	✓	×
C.	Drenches (after germination/transplanting)		
	1. Dexon	×	✓
	2. Caton	×	✓
	3. Ferbam	×	✓
	4. Thiram	×	✓
	5. Zineb	×	✓
	6.Chloronob	×	✓
	7. Bordeaux	×	✓
D.	Dust (After transplanting)		
	1. Mercury	✓	×
	2. Captan	✓	×

Note: × stands for No.
✓ stands for Yes.

An alternative is transplanting seedlings directly from germination trays into the pots, but it is not generally followed mainly for the following two reasons, viz.

1. The germination trays are usually filled with heavy soil, making the trays difficult to lift easily and handle conveniently under such operations; and
2. Wet seedlings root plant better than those coming directly from the germination trays.

(b) Re-sowing and watering: In direct seedling, if, germination fails to a great extent or unsatisfactory, then resowing is carried out in the contains. The regrouping of pots before the operation of resowing prevents the suppression of young germinating seedlings. If, some voids appear in some beds, then they should be filled soon after with extra seedbed seedlings. Adequate amount of moisture is essential during the process of germination. It has been advised sincerely, that the germination trays must be checked two or three times a day. The over-head spray prevents the washing seeds out of the trays (Fishwick 1966). Here, one very important precaution must be taken that, if over-watering occurs, then shade must be removed for the time-being. This will help to speed up the evaporation.

(c) **Shade and weeding:** They are generally shaded with the following very important objectives, viz.,

- **Protection against sun:** The young seedlings of shade-bearing delicate tree species of Lauraceous and magnoleaceae families should be protected against the scoching sun when the seeds are shown during the months of February and March (before the monsoon sets in).
- **Protection against frost:** The seedlings of frost-tender tree species require protection against the morning frost-oriented locality. Therefore, the shade is usually removed almost daily as the temperature begins to rise quite high. If, it is not done then the soils may get chilled due to non-availability of sunlight. This may result in greater damage to the seedlings.
- **Protection against rains:** The seedlings are protected against the impact of rains and hails. This is carried out in spring and autumn seasons. The falling rain drops expose the roots of seedlings. The seedlings are covered with splashed mud and may get killed there itself. In the same way, the hail stones may kill the plants/seedlings by causing physical injuries.

The shades are generally made up of thatch, mats, split bamboos, battens, polythene or tin sheets. The shades are used, the plants get used to them and, hence, it is urgent to harden-off the seedlings before the operations of transplanting or even the planting out.

After the seeds germinate, there may crop up the weeds also which usually compete with the seedlings for food (nutrition) and light. Therefore, the weeding is necessary to get rid-off the weeds (unwanted plants). The negligence and the delay in such operations may result in the killing of the desired seedlings. Hence, the nurseries should always be kept free of weeds.

In case of development of tap-roots and the lateral roots, they are usually pulled out after the loosening of soil by point iron nails/or a pointed stick made up of iron.

Precaution: In this operation, the following few precautions are very essential and urgent:

(a) The weeding operation must be got done under strict supervision;

(b) The roots of the seedling of desired species should be disturbed during weeding operation;

(c) The weeding should not be carried out when the soils are wet because there may be always a danger of uprooting of the desired seedlings.

Operations: While weeding's are carried out, the following two operations are also performed:

1. The upper surface of bed should be worked up with the suitable implements of weedings for reducing the evaporation of moisture; and

2. The seedlings of the desired seedlings of species should be well-thinned-out, if, they become crowded under conditions.

2. Bare-root system

(a) **Shading:** Lath shade can be provided for,

- Small round-level bare-root beds; and
- Raised bare-root beds.

The transplanted seedlings should be placed under at least 30% shade for one or two weeks. However, some areas use up to 85% shade (how craft 1973). Because the seedlings suffer the double shock of removed from shaded germination beds and the planting in the exposed containers, the risk of wilting and death is too great.

For raised transplant beds, the shade screens are placed high enough above the pots, so that the tending of seedlings may be carried out in an easy way. Some other shade materials are the bamboo (Kana) screens, palm fronds, and the wooden laths.

(b) **Watering:** It is evident that the bare-root seedlings are more usually exposed to air and light than that of the container stock. Baur (1960) advocated the water needs of unscreened bare-root seedling stocks are much greater than those of screened container seedling stock. The nursery beds are checked frequently wilting. It has been reported elsewhere that approximately 100 liters of water is required per day for a bare-root bed of 1.2 × 5.0m size.

However, the actual water requirements vary by sites and are just related to the following:

- Bed soil texture;
- Amount of solar radiation received;
- Presence/or absence of winds;
- Species transpiration rates; and
- Growth stage of seedlings.

McDonald (1984) said that for medium and large bare-root nurseries, the fixed irrigation systems give very promising result by best water coverage. However for small nurseries, water horse and sprinkling cars can be suffice. The sprinkler irrigation system is quite useful as the water is spread on the seedling in a uniform pattern.

(c) **Weeding:** In bare-root seedlings, hand weedings is practical when the beds are small. In such case, labour should be,

- Plentiful is number; and
- It should be cheap.

However, when the seasonal production in bare-root nurseries reaches upto 1,00,000 to 2,00,000 plants, the chemical control is more practical and economical.

The major concern is selecting the best and suitable herbicides for weeds that are just non-toxic to the growing seedlings.

Audus (1976) reported that technical classification use chemical structures and major functional groups at the highest level of categorization. A list of suitable herbicides is given as under:

	Herbicide group	Specific herbicide example
1.	Phenoxy compounds	2, 4-D; 2, 4, 5-T; MCPA; Silvex.
2.	Benzoic acids	Chloramben (Amiben)
3.	Pichoram	Amolon
4.	Benzonitriles	Bromoxynil
5.	Bipridinium salts	Pareqat; diqat
6.	Amitrole	Amerol; cytrol
7.	Aliphatic acids	Dalapon; TCA
8.	Substituted phenols	Dinoseb; DNBP
9.	Meleichydrozide	MH
10.	Substituted areas	Chloroxuron; Diuron; Linuron
11.	Triazines	Simazine, atrazine, prometryn.
12.	Diazines	Pyrazon
13.	Amides	Diphenamid; propanil
14.	Carbamates	Chlorpropham; CDEC
15.	Trifleoralin	Persan; Treflan
16.	Xylene	Dimethyl benzene

- **Poor weed control:** This problem is due to the failure to use selective herbicides for some weeds and failure to apply herbicides at correct rates, right light or in the best manner possible.
- **Seedling injury:** The herbicides are often blamed for injuries that are also caused by some insects, diseases, winds, over-watering, nutrient deficiencies, and abnormal weather conditions.

LATER TENDING

The later tending comprises of the time from four to six weeks after germination until out-planting covering the lifting operations. The period is just different from the period of early tending. Some kinds of problems such as weeds, moisture and lifting still remain there. These factors are urgently needed to address that may promote proper seedling root and shoot growth.

1. Container system

It is expected that the nurseries should produce very healthy and sound (bold) seedlings. There should be minimal differences in the seedling regarding the following characteristics:

(a) Heights;

(b) Root collar diameter; and

(c) The foliage.

The grouping seedlings in beds by height, produces the planting shock of uniform size (i.e., the height of the seedlings). This is the case of container systems of seedling production.

However, in bare-root systems, such type of sorting is not possible and uniform size can be achieved by some other means, for example, root-pruning, under-cutting, chipping, etc. In such case, healthy, well-developed root-system will help such seedlings to maintain vigorous growth in the nursery beds and can adapt very quickly to the new areas and the environment after the operation of out-planting.

The potted seedlings can be lined-out on the ground and may also be put on the other surfaces. This operation may be carried out in the permanent raised or may be in the second beds. The air-pruning of roots may be performed. However, if, the roots egress into the underneath soils, the lifting and packing may become very difficult and there is a danger of root damages. Therefore, every precaution in lining-out must be taken care of.

Mycorrhizal fungi: They may help in mineral as well as water uptake. This may be most probably by increasing the root-surface areas. It is oftenly advised that all coniferous seedlings should be inoculated very carefully in the nursery. Also, mycorrhizal fungi can be added to the pot medium. This can be accomplished just before transplanting.

The mycorrhizal fungi should be get locally. Mostly these are collected in mineral soils and also in the duff. These fungi should be collected from a health forest stand. All the material should be grounded and put into the pot medium.

Watering and weeding: Water is very essential for living things. The tree species seedlings should be watered as and when required. This depends upon the experience of nurserymen. Here, the practical knowledge of the following factors is of utmost importance, viz.,

- Plant stress;
- Different pot volumes;
- The medium used in the pots;
- Time since the last irrigation/rainfall;
- Current condition of the moisture of potting medium; and
- Growth seedling stage.

All over the world, the early morning watering is supposed to be the best. Afternoon irrigation/watering may reduce the chance of diseases. As the seedlings get growth in good dimension, and the root systems expand in the pot medium, the moisture stains can be determined and estimated just depending on the following few but very important factors, viz.,

- Volume of the container;
- Type and texture of growing medium; and
- Ability of pots to void excess water.

However, over-watering/excess watering saturates the medium and thus the aeration is reduced drastically which may prove fatal to the proper growth and development of the seedlings. The growth and development of the seedlings. The results of poor drainage and aeration are,

- Stunted growth;
- Yellow coloured seedlings foliage; and
- More concentrated root growth of seedlings between the soil and bag interface, them in the pot medium.

If, nutrients are added through irrigation water, then salinity and acidity of the soil medium must be looked into regularly for maintaining the same. In case of hardening-off, the watering should be gradually enhanced with with a great frequency.

The weeding is a crucial factor in container system. When the seedlings grow after germination increasing foliage area shades the pot surfaces and this helps to suppress the growth of weeds. The weeds must be controlled at all costs to save the seedlings. Hand weeding is the most common practice under nursery conditions. The unchecked weed growth in these areas may cover the nearby seedlings by shading them. The shovels, hoes and some other small hand tools are usually used for weeding out the grasses and unwanted plants (Weeds) from the nursery beds.

Controlling shoot and root growth: The shoot growth is controlled by grouping the seedlings of similar height mainly in the same bed. This method may help in keeping the fast growing seedlings from shading out the slow-growing ones. Moreover, the very fast-growing seedlings may be clipped out to check the growth.

The root growth of potted seedlings in just controlled by effective pot volume and depth. If, labour is somewhat cheap and in sufficient in numbers in the area, the seedlings may be re-potted in larger containers. However when re-potting is impossible, several root-pruning is just required before out-planting process.

Transporting the seedlings: The containers are put in the transport flats and loaded on to the tracks tractors. The wind burn is a sever threat to the seedlings. Therefore, the seedlings must be saved from wind and the scorching sun. however, in some areas, some anti-transpirants, if, possible, must be sprayed on the foliage of seedlings before they are transported (in case of temperate climatic conditions). But in tropical climates, the use of anti-transpirants is not generally necessary for seedlings because they are not stored in dormant conditions.

2. Bare-root System

The mycorrhizal fungi can be used as inoculation of seedlings under bare-root nursery beds. In this case, the forest duff and mineral soil or/water-born spores are used for container beds. The inoculation is carried out before sowings/or before seedlings are four to six weeks old.

Watering and weeding: In a general sense, the bare-root beds have greater water requirements than that of the container beds. This is because of their greater environmental exposure. It is also evident that older bare-root seedlings have greater water requirements than that of the young seedlings. In this case, every day watering is need until the time of hardening-off starts. Soil moisture content must be checked regularly in well-drained beds of nursery. It has been noticed damaging effects of irrigation droplets impact are minimal for older and larger seedlings.

The openness of the bare-root beds generally frame the good dias for invitation to the weed growth. Further, hand weeding is practiced in bare-root beds. However for larger nurseries, only the use of herbicides is very effective. The application of herbicides last longer (about 30 days). Also, sometimes kerosene/or other mineral spirits control the weeds to a great extent in bare-root nursery beds.

Shoot/root growth control: The following are the factors/techniques to control the root and shoot growth of bare-root nursery seedling stock:

- Water cutting;
- Lateral pruning; and
- Top pruning.

In small nurseries, none of these techniques may be used. The reason could be the scale of operation does not demand quality control of seedlings stock.

- The under-cutting is performed with shovels in the small nurseries whereas in large-sized nurseries tractor drawn machines are used. The technique, however, is most effective in just manipulating the shoot/root ratios (shoot-root).
- The lateral root pruning serves the roots between adjacent the seedling rows. This technique may allow the easier lifting of seedlings. This is oftenly done before the process of packing and transporting. However, in this technique the machines are used, may generally depend upon the size of nursery.
- Top-pruning, which is also known as clipping, generally controls the growth of shoot. This technique is mostly used for individual seedling under the bare-root systems. The operation enhances the root growth and hand shears are used in small nurseries whereas the tractors are used for large nurseries. This depends upon the dimensions of the bare-root system.

Transplanting: There are generally four general procedures for carrying the bare-root seedlings to the field are as following:

- Lifting;
- Protecting;
- Grading; and
- Transporting.

The seedlings should not be pulled out directly from the seedbed. The soils is shaken-off the roots, which are just washed, trimmed, and then kept in water, or water slurries, or wet peat moss. If, the plants/seedlings are too large in size, prun them back to a total length of 18 to 30 may improve the survival of the out-plant. There is no problem, if, the lifting and transporting is done on the same day or within the few hours of lifting from the bare-root system.

Stump planting: Jackson (1974) indicated that there are some tree species, e.g., *Cassia fistula*, *Gmelina arborea*, and *Tectona grandis*, withstand the trimming of the major portions of the seedlings stems and root system. This results in a stump having the length of 15 to 30 cm. The shade and protection measures are the same as that for bare-root stock.

Other Consideration

1. **Shoot/root ratios:** The seedling health and suitability for out-planting is assessed by the shoot/root ratios. This ratio is calculated just by measuring the oven dry weight/volume of seedlings tops divided by oven-dry weight/volume of seedlings roots. However, the shoot-to-root dry weight ratios of 2:1 are most acceptable for several tree species.

   However, the higher ratios generally indicate excessive shoot growth for just corresponding root growth. The greater root branching is usually in the medium with peat, vermiculite, and perlite than in the soil only. In case of soil medium only, there could be the greater development of top roots and less lateral branching which may cause root curling/or spiraling at the pot bottoms. However, large spreading root-systems are supposed to be the best due to the following two reasons:

   (a) Their greater capacity for absorbing water and the nutrients from the soil/ or may be for pot medium;

   (b) Greater surface area from which new lateral roots come up just after out-planting.

2. **Hardening-off:** With a great care of seedling and the treatment thereof, permits the production of green, healthy and vigours seedlings which are most suitable for out-planting. Therefore, for getting large seedlings with vigorous growth in a very short periods, the following are used/should be used frequently:

   (a) Water-appropriate quantity;

   (b) Fertilizer-optimal does serves the purpose; and

   (c) Shade-good shading is also essential.

   The hardening-off process increases the chances of seedlings survival. This process, however, conditions both the bare-root grown seedlings and containerized stock. The seedlings can survive longer period without water and shade even. The hardening-off the seedlings in container beds, there exists a relationships between the following:

   (a) Pot volume;

   (b) Potting medium;

   (c) Seedling size; and hardening-off must be considered very carefully.

3. **Culling J-root seedlings:** The hardwood seedlings show some mal-formed top-roots in the nursery conditions. The J-roots may occur in the process of transplanting from the germination trays or the nursery beds. This poses a serious problem because the J-shaped rooted plants do not grow in the shape of healthy, good, and straight plants. It is, therefore, always better to transplant the seedlings before lateral root branching occurs and before the seedling flush takes placed in the nursery beds.

Chapter 9

PROTECTION OF NURSERY

INTRODUCTION

The protection of nursery is the prime aim of getting good seedling stock. In general, the stock of nursery requires a large investment in terms of the following:

1. Time;
2. Labour; and
3. Monetary resources.

Cordell and Filer (1984) and Sutherland (1984) strongly advocated that without protecting the above mentioned investment, a large seedling loss may occur which is a great setback to the growing of nursery of any type. Therefore, the large and small sized nurseries require protection from many physical and environmental elements and side by side from the insect-pets and the diseases. It has been reported somewhere in forestry literature that the insect-pests and diseases cause a great nuisance to the nursery seedling stock. It is a very costly affair to protect the seedlings as well as the property. In some countries communities, the payment to the nursery workers in terms of food than the direct cash may be the best option. However, now in the changing environment, the man has become materialistic in nature and hence instant in cash rather than any other options.

In nurseries, sufficient fences deter the animals and people. Therefore, the fences made of treated wood and with a proper materials may last long. When diesel and pentachlorophenol are available locally and at somewhat cheap rates, the hot-cold treating of posts (fence posts) in some large pots/utensils is somewhat inexpensive. However, in some cases, the live-fences and hedges mixed with thorny type of plants/ or some local shrubs may be sufficient in many respects.

It has been experienced that the windbreaks may reduce dry winds and its effects that usually evaporate moisture from the seedbeds (Kittredge 1973). For this purpose, local tree species are generally preferred, but prolific seeders like, sububool (*Leucaenal eucocephala*), black locust (*Robiniapsoud acacia*), etc., are generally discarded. It is a hard fact that wind blown tree seeds from the windbreaks may get germinated and may compete with seedling root system which grow in seedbeds. Moreover, the removal of these invaders is somewhat may get disturbed in the weeding operations under nursery conditions.

The forest fires are the great havocs which may destroy the entire nursery, plantation and even the forest wealth within no time. The fires work both as friend and enemy.

Weber (1977) further reported that where the fires are frequent, a 2.0 to 3.0m wide fire break must be maintained around the whole nursery area/site. Water outlets should be installed at several places very near to the perimeter. It will certainly make sure that water is readily available, if, a fire, after outbreak, jumps the fire break.

1. **Insect-pests:** Diversity of climate and environments in the tropics pre-clouds simple prescription for insect problems of forest nurseries. The nurseries have become very important regarding the supply of healthy and tall plants for plantation purposes. It is, therefore, very essential to protect nurseries from insect-pests. The following insect-pests of forest nurseries may seriously affect the supply of seedlings/plants:

   (a) Cockchafers: These are the root-feeders and just kill the seedlings by damaging the rootlets/or removal of bark of top-roots. They lay eggs in the soil. The larva is a white grub upto 2.5 cm long and generally curved in the shape of 'C'. The following applied to control the pests:

   Silvicultural:

   - The beds should not be prepared in the hot season, or during the early part of the rainy season because this is supposed to be the egg laying period.
   - The soil working has to be carried out in the oviposition period, the worked up soil should invariably be covered by a layer of charcoal, sand, or sulphur.
   - In transplants beds which are not to be utilized immediately they should be water with kerosene oil emulsions.

   Mechanical: At the time of bed preparation, the white grubs should be picked up and then destroyed by some mechanical means. It has been noticed that grubs leave the soil at the time of nighty, and travel on the ground surface. Such species of insects can easily be trapped by the help of sinking tins/or earthen pots flush with the ground between the beds.

   Chemical: The following chemicals can easily be used to control the insect-pests:

   - CS_2;
   - CaCN;
   - DDT;
   - B.H.C;
   - Nepthalene;
   - Aldrin dust; and
   - Acetic acid.

   (b) Cut worms: The cut worms/or surface caterpillars damage the nursery as they cut-off the seedling at the ground level. The conifers (Conifer tree species) are the most victims of cut worms because the conifers do not coppice. The following are the control measures:

- **Silvicutltural:** The nursery beds is well as the nursery sites must be fore from the weeds the grasses. The young larvae that usually feed on the weeds, die automatically for want of food material.
- **Mechanical:** The usual practice is just to collect the larvae by hand and hill them on the spot. Further, the succulent leaves of tree species, weeds, and grasses are heaped up at one place. The cutworms are attracted towards this heap for feeding purposes and the insects can be killed on the next day.
- **Chemical:** A poisonous bait may be made/prepared by mixing with bran (2.5 parts) and sodium fluoride or lead arsenate (1/6 part) dry and then in a mixture of molasses (1/4 part) and the water (4 parts by weight). This mixture can be spread # 2.25kg/100m^2, just very near to the affected area shortly after the sunset.

(c) Cricket: The insect are very injurious to the nursery. They feed on young seedlings and low shoot cutting them at night. Following few measures of control are;

- **Flooding:** The tunnels should be detected and water or water with kerosene oil mixed in it. This mixture is poured on them in the tunnels.
- **Trapping:** The insects can be trapped by sinuating carthen pots in the beds which is half-filled with water and some amount of kerosene oil mixed in it.
- **Baiting:** Just half of the pumpkin is placed in the nursery beds which is covered with weeds in the evening. The crickets take shelter in the and they be killed and sometimes some poisonous baits can be tried to kill these insects.

(d) Grass hoppers: These are also injurious to the seedlings in the nursery beds. Handpicking method is employed to kill these insects. The foliage spray of 0.5% aldrin/malathion gives very promising results.

Common sucking pests: Following are the very common sucking pets of nursery:

1. Aphids;
2. Thrips;
3. Mealy bugs; and
4. Scales and mites.

Some larger pests:

1. Spiders;
2. Birds; and
3. Rats and mice.

An insecticide, e.g., spectracide may be strayed to kill the spiders that generally damage the forest tree seedlings. A flock of birds can easily eat several thousands seeds per day. The repellents (with stickers) are available to control feeding on the seeds by birds.

Neither sticker nor repellent should interfere with germination. A summary of major pests, species attacked, symptoms, and recommended treatments is given as under (adopted from liegel and venator, 1987):

	Pests type	Seedlings species affected	Control
A.	Chewing mouth parts		Ingested poisons
	1. Cut-worms	Mahogany, pines, teak	Spectracide, chord
	2. Crickets	Teak	General insect wide
	3. Leaf miners	Mahogany	General insecticide
	4. Shoot borers	Mahogany and other meliacae family species.	Systemic insecticides
	5. Sugarcane root stack borers.	Mahogany	Soil drench insecticides
B.	Piercing sucking mouth parts:		Contact poisons
	1. Aphids	Teak, kadam, grnelina	Various
	2. Mealy bugs	Teak	Various
	3. Thrips	Kadam	Varous
	4. Scales mites	Pines	Various
C.	Large pests		Various controls
	1. Spiders	Pines	Ingested/contact poisons
	2. Birds	Pines/hard woods	Chemical repellent added to seeds
	3. rats/mites	Pines/hardwoods	Chemical repellents cover seeds; pet cut in nursery

2. **Diseases:** Somewhere in forestry literature, it has been mentioned that the disease symptoms can include sooty appearance, a form of fungus, rust spots, irregular-shaped blotches, or the mosaic of different colours, and sticky discharges. The discolouration symptoms can be caused by either disease organisms or by some mineral deficiencies. A summary of major diseases with some characteristics are given as under (adopted from Liegel and Venator, 1987):

Diseases	Seedling species affected	Control measures
A. Damping-off fungi	Pines/hardwoods	Pot media and sterilization; soil drench fungicides.
B. Seedling needle blight	Pines	Various fungicides.
C. Brown needle disease	Pines	Bordeaux mixtures (i.e. copper sulphate)
D. Root rot fungi	Pines/hardwoods	Soil sterilization good watering control and monitoring

The diseases occur in the nurseries and knowledge of the nursery disease is very essential.

CLASSIFICATION OF NURSERY DISEASES

1. Common diseases;
2. Diseases of conifer;
3. Diseases of broad leaved species; and
4. Diseases of exotics.

Now a brief description of each class of diseases is discussed below.

Common Diseases

1. **Damping-off:** It is the most serious of all the nursery diseases. It is primarily a disease of conifer seedlings, but, it attacks broad-leaved seedlings also. The damages caused by this disease is that it can wipe out the whole nursery stocks. The mortality of seedlings occurs at three stages of growth of seedlings:
   (a) The seedlings may be killed after the time of emergence from the ground as a result of rotting of tissues at the base;
   (b) The seedlings may be killed even before they emerge above the ground level due to pre-emergence blight;
   (c) The older seedlings may be killed/die due to the root-rot.

   The damping-off disease is caused by a number of fundi, the most important of which are as follows:
   (a) Pythium;
   (b) Rhizoctonia;
   (c) Phytoshthora; and
   (d) Fusarium.

   In India, rhizoctoniasalani is the most important cause of damping-off in the nurseries.

How to control?

This disease can be controlled and prevented by the following measures as discussed below.

1. The nursery must be situated in a well-drained soils, light textured soil without my overhead shade;
2. The soil may be partially sterilized by means of steam/or fumigants;
3. The soil in the seed containers may be sterilized by keeping the seed containers in an auto-cleave and then allowing the steam to run-over it;
4. Minimum watering must be done as this disease is more prominent in the damp conditions;
5. If, the soil of the seed beds is to be sterilized, steam pipes bored with the holes are put inside the pipe and the steam is passed through it for a period of 30 minutes;

6. The nursery beds may also be sterilized by the application of some chemicals like, formmaline, methyl bromide, etc. At least the application of these chemicals should be carried out for about 2 weeks before sowing operations as these chemicals are highly phytotoxic in nature; and
7. The treatment of soil and the seed with chemicals before the sowing in the beds, is another very important and effective method for controlling the damping-off. However, some of the chemical products recommended for controlling damping-off disease (Reddy and Mishra, 1970):

Chemical	**Dose/sq. m of bed (g)**
1. Captan	24.5
2. Thiride	26.9
3. Cuman	26.9

Source: Khanna, L.S. (1982), Forest protection, KhannaBandhu, Dehradun (Uttarakhand), pp. 206.

Therefore, for protecting the conifer seedlings against damping-off, only soil treatment should be followed. This will give a great relief in controlling the disease to satisfactory level.

Diseases of Conifers

1. Abies pin draw (Fir): Piceasmithiana (spruce)

1. Root-rot and mortality is caused by armillariamellea;
2. A normal damage is also caused by fomesannosu;
3. Heart rot in fir and spruce is usually caused by several heart rot fungi, e.g.,

Fumes	Fomentarius
F.	robustus
F.	Pini; etc.

4. The red colouration of tree wood is caused by a fungus called fusaricum mostly in spruce (*Picea smithiana*);
5. Needle rest is another common disease of both fir (*Abie spindrow*) and spruce (*picea smithiana*) which is caused by *Peridermium thomsoni*; P. piecea; and cerapsorapiceae.

2. Cedrusdeodara (Deodar)

Fomesammasus is a widely spread fungus which causes,

1. Mortality in young deodar.
2. Butt rot in somewhat older trees.
3. Reduction in growth due to death of roots.

Control measures: The disease is generally controlled by the following measures:

Silvicultura

1. The fungus sustains itself as a saprophyte on the dead roots, therefore, the only possible control may be to dig out the roots completely before raising plantations; and
2. Well-drained site/areas should be selected for growing plantations.

   Chemical: The spread of disease can be prevented by some chemical treatment of freshly cut stumps. However, the chemicals usually which are used may be creosote, urea (20%), or sodium nitrite.

   Biological: The inoculation of *Peniphora gigantean* acts as a detergent against the disease fomesannasus

   Some other diseases of deodar (Cedrusdeodara) may be caused by:

   (a) Peniphoraluma-butt rot and trunk rot;

   (b) Polyporustomentosus-root rot and butt rot;

   (c) Peridermiumcedri-witches broom.

3. Pinusroxburghiisargent (Chir pine)

Cronartiamhimalaynese causes swertiafeltrust;

Besides this disease, fomespini, E. piniculata, ganodermaapplanattum cause heart-rot in chirpine (pinusroxburghiisargent); and poria monticule causes cuboidal decay.

4. Pinuswallichiana (Blue pine)

Fomespini, commonly known as trametespini causes decay in heart wood of pinuswallichians (blue pibne). The fungus attacks the trees mainly through the wounds in the stem. Therefore, lopping of blue pine (pinuswallichiana) must be banned.

Diseases of Broadleaved Species

1. Acacia catechu (Khair)

Ganoderm alucidum causes root-rot.

1. Young plants are likely to die-off in a very short time after the infection; and
2. Mortality of mature trees occurs only when a large portion of its roots are affected or ultimately gets killed there in.

Fomesbadius causes heart rot in khair through the injuries on the stem.

1. Digging isolation trenches to just isolate the infection of disease.
2. By digging out the affected roots.
3. The species resistant to this fungus may be mixed with plantations of khair (*Acacia catechu*). The resistant species are identified as follows

   Bombaxceiba, Alianthus excels, etc.

2. *Dalbergia sissooroxib*, (Shisham, tali)

Fusarium solani causes wilt disease that usually results in yellowing and drying up of leaves. After the foliage is wiped out completely, mortality may occur within a very short time.

The control measures include raising of shisham (*Dalbergia sissoo*) plantations on well-drained light textured soils. However, root-rot is caused in shisham (*Dalbergia sissoo*) by *Ganoderma lucidum*.

3. Santalum album (Sandal wood)

Sandal wood is a very precious and valuable timber of South India where it is found in abundance in the states of Karnataka and Tamil Nadu. The sandal wood has been affected by an epidemic of a highly dangerous disease usually known as 'spike disease'. The resistance of a tree to this disease increases with the age and there could be very less chances of a tree being affected one it reaches the age of 15 to 20 years. The disease, in the part, was attributed mainly to be due to the following factors (Khanna 1982):

1. Intensified carbon assimilation;
2. Death of haustoria and the root-tips cutting of supplies of salts and water;
3. Attack by virus; or
4. Imbalanced circulation of sap.
5. Micro-plamsa like bodies (MLBs) are also considered to cause 'sprite diseases' in sandal wood (*Santalum albumn*)

Control Measures

1. **Sanitation:** For preventing the disease of sandal wood, the diseased plants are removed, thus, the spread of the disease may be lessened to a great extent. The spider sandal trees are removed at regular pace. But this is not a good practice as many plants are removed without reducing the incidence of disease of sandal wood.
2. **Selection of hosts:** The efforts must be made to raise sandal wood with hosts which may import resistance, for example, *Azadirachta indica*, *Cassia siamea*, *Dalbergia sissoo*, etc., and not with the hosts of *Acacia spp*., *Cajanus cajan*, lantana, etc. which just make it susceptible to the disease. Since the sandal wood (*Santalum album*) is a light demander, so the host plant should not be only fast-growing but have light crown also.
3. **Vector host plant:** Sandal grows very healthy in nature under a top canopy of the hosts like that of azadirachta, eassia, casuarinanad eucalyptus, it is possible that freedom from the disease may be just due to the canopy. However, Bakshi (undated) suggested the control of disease by raising sandal wood under light canopy.

4. A few efforts have been made to control this disease of sandal wood by using some chemicals. Hence, tetracycline, and other antibiotics provide relief for a very short period.

4. *Shorea rubusta* (Sal)

1. *Polyporuss horeas* causes root-rot in high rainfall areas.
2. Heart rot is caused by the following:

   Hymenoclaceterubiginosa, fumes caryophylli, and fomesfastuosus. Hymunochaeterubiginata attacks the trees through the wounds generally caused by the fire.

Control measures: Sal diseases may be controlled by the following:

1. Preventing injuries to the forest;
2. Damage by severe hot weather can be reduced to a great extent by early burning operations;
3. Regular thinning and improvement fellings should be exercised at regular intervals;
4. Girdling of stools and then applying the poisons like 2,4-D or 2,4,5-T, etc.
5. The mortality in Sal (Shorearobusta) is due to site deterioration, it can be controlled by adopting the following:
   (a) Fire protection;
   (b) Closure to grazing;
   (c) Contour bonding, etc.

5. *Tectona grandis* (Teak)

Pseudomonas solanacearum causes the wilt disease not only in seedling but also old transplants in *Tectona grandis* (Teak).

Peniophorarhizomorpha-sulphurea and Polyporus zonalis cause root rot in India and may generally result in wind-throw of affected trees of Tectona grainds.

Also, corticium salmonicolour causes the pink disease in teak (*Tectona grandis*) in high rainfall areas of Kerala and Karnataka state but it has been noticed that damage is not so serious.

Control measure: The following control measures are taken into consideration:

1. Raise plantations on very good sites;
2. Soil must be sterilized before use;
3. Watering may be done sparingly;
4. Weeding must be carried out a regular pace;
5. Use of chemical fertilizers instead of the farm yard manures;
6. Minimizing injury to the trees by suitable methods; and
7. Only very healthy stems should be planted in the nursery areas.

Disease of Exotics

1. *Casuarina equisetifolia* (Saru)

Trichosporium vesiculosum causes the stem wilt in *Casuarina equisetifolia*. Here, the pathogens of the disease at initial stages enters a healthy plant in a new plantation by injury caused during the following functions, viz.,

1. Pruning; and
2. Illicit lopping of branches.

Once the roots are infected, the disease spreads by root contact and root grafting. But, if, the disease has already entered in a plantation, then its spread can be checked by making trenches round the diseased tree which generally occur in to groups. Further, the dead trees should be removed as soon as possible to just remove the source of infection.

2. Eucalyptus hybrid (Safeda)

This exotic species is being planted on a massive scale in most of the India states since 1962 whereas eucalyptus globules was introduced in Nilgiris somewhere in the nineteenth century and it has naturalized there itself. The most profound disease to which this widely planted exotic is exposed is the pink disease which is caused by corticiumsalmonicolour.

However, the fungus is found naturally in India and attacks the exotic at a very early age. This results in repeated die-back and mortality which can go upto cent per cent in worst damaged and affected plantations. The escape infection in early life, may also get affected later on. This may result in the development of cankers.

The eucalyptus also suffers from root rot disease caused by ganodermaucidum which results in wind-throw in plantations. In terai areas of Uttarakhand state, eucalyptus hybrid generally suffers from the following characteristics:

1. Yellowing of foliage
2. Die-back
3. Pre-mature defoliation; and
4. Eventually the death of seedling/plants.

Chapter
10

NUTRITION MANAGEMENT

INTRODUCTION

The performance and growth of nursery stock is quite related to the nutrition of seedbed soil or the container medium. The fertility of soil is estimated by both chemical and biological methods. The deficiency, if, any can be corrected by adding organic or inorganic fertilizers.

ESSENTIAL ELEMENTS

Tisdale and Nelson (1975) reported that there is disagreement on which elements are truly essential for plant growth. However, carbon, hydrogen and oxygen comprise the bulk of seedling weight. The elements are traditionally called macro-elements or micro-elements, depending upon whether plants use greater or lesser quantities. Epstein (1973) advocated that elemental concentration may vary by plant component (leaves, twigs, branches, fruits and the roots), age, position within the seedlings (top, center, or lower stem area), season, and climate (dry or wet). Known functions of each element are as follows:

Elements	Functions in Seedlings Nutrition
	MACRONUTRIENTS
1. N (Nitrogen)	Essential part of chlorophyll and proteins, fosters the vegetative growth.
2. P (Phosphorus)	Promotes root growth and stem strength; important in all energy transfer within the plants.
3. K (Potassium)	It promotes starch and sugar formation, root growth, resistance, shoot strength and plant vigour.
4. Ca (Calcium)	Enhances N uptake, cell elongation, and development of meristematic tissues.
5. Mg (Magnesium)	Major constituent of chlorophyll in plants; important in P metabolism and enzyme activation.
6. S (Sulphur)	Essential for synthesis of S-containing amino acids and nitrogen fixation by legumes.
	MICRONUTRIENTS
7. Fe (Iron)	Important in chlorophyll production and enzyme activation.
8. Mn (Manganese)	It activates the enzymes regulating carbohydrate metabolism and essential for photochemical (Hill) reaction.

9. Cu (Copper)	Metal activator for several enzymes and possibly light reaction in plants.
10. Zn (Zinc)	Metal activator of enzymes.
11. B (Boron)	Responsible for several regulator mechanisms in many plants, including sugar translocation across cell membranes and the oxidation processes.
12. Mo. (Molybdenum)	Required for N-fixation in the legumes and nitrate reduction non-legumes.

Source: 1. Leonard, Dave (1980) Soils, crops and fertilizers use or a what, how, and why guide, 3rd ed., information collection, and exchange. Reprint R8, Washington DC; Peace crops. 162 p.

2. Tisdale, Samuel L; Nelson, Werner L. (1975), Soil fertility and fertilizers. New York: Mac Million 694 p.

Deficiency Symptoms

The very common symptoms are discussed below.

1. Severe stunting conditions;
2. Delayed or abnormal growth;
3. Leaf discolouration (bright yellow, purple, or red instead of green); and
4. Abnormal root growth.

Lack of one element may cause divergent symptoms in some species. However, similar type of symptoms are caused by deficiency of several elements, for example general type of chlorosis by lack of either nitrogen (N) or sulphur (S).

There are some symptoms which are caused by multiple nutrient deficiencies whereas others are usually not nutrient related at all. However, whenever the deficiency symptoms are seen, some damage of irreparable nature may have occurred in shoot or root tissues of the plants. A generalized 'key' to the visual symptoms are as under:

Symptoms	Symptoms may be related to
1. The dominant symptom is chlorofic foliage (Discolouration)	
(a) Entire leaf blades are chloric	
• Only the lower leaves are chlorotic followed by necrosis (tissue death) and leaf drop.	Nitrogen
• Leaves on all parts of plant are affected and often have a beige cost.	Sulphur
(b) Yellowing of leaves takes the form of intervened chlorosis.	
• Only older leaves exhibit inter-veinal chlorosis.	Magnesium
• Only younger leaves exhibit inter-veinal chlorosis	Iron and/or manganese
(*i*) This is only symptom;	
(*ii*) While younger leaves have inter-veinal chlorosis; the tips and the lobes of leaves remain green, followed by veinal chlorosis and rapid extensive necrosis of leaf blade.	Copper
(*iii*) Young leaves are very small, sometimes missing leaf blades altogether, and internodes are short, giving a rosette appearance.	Zinc
2. Leaf chlorosiss is not the dominant symptom.	
(a) Symptoms appear at the base of plant.	
• At first, leaves are dark green and smaller than normal. Later, purple fragment develops in older leaves.	Phosphours
• Margins of older leaves burn or small necrotic spots appear on older leaf blades.	Potassium
(b) Symptoms appear at top of plant.	
• Terminal buds die, giving rise to a witches-broom.	Boron
• Margins of young leaves fail to form, sometimes yielding trap-leaves. Growing point ceases to develop, leaving a blunt end.	Calcium

Source: Adapted from material of P. Nelson, Horticulture Department, North Caralina State University. Raleigh (U.S.A.)

Foliage (Tissue) Sampling

Many workers like, Minko and Craig (1976); Simpson (1985) and Youngberg (1984) advocated that foliage, tissue or plant analysis is generally used operationally to determine seedling nutritional before deficiency symptoms appear. The excessive amounts of some elements, e.g., Fe (Iron), B (Boron, or Mn (Manganese) are toxic to the plants. However, the plants absorbs excess nutrients without harmful effects.

Luxury Consumption

Epstein (1973) narrated that good plant growth with nutrient levels beyond these normally needed. It has been noticed that in temperature regions, the foliage is usually sampled at the time of first hardening, in late summer because the nutrients concentrations are somewhat less variable. The very promising results can be got just be sampling:

1. Mature;
2. Fully developed leaves; and
3. Fully elongated needles.

The collections are generally carried out in the early hours of morning. Before this the seedlings begin synthesizing CHO (Carbohydrates) and utilize the nitrates accumulated at night. The foliage analysis, interpretations should consider the following:

1. Environment of the seedlings tested; and
2. Their physiological state.

The foliage analysis shows two disadvantages viz.,

1. The collaboration trials for setting high, low, medium, and critical levels for each element are,
   (a) Very costly; and
   (b) Time consuming.
2. The techniques and equipment needed for chemical analysis are rather very sophisticated, are require highly trained personnel.

Further, if, some facilities exist for agriculture foliage research, then the techniques may easily be adopted for forest nursery diagnostic work.

Soil Tests

It is believed that soil test are generally superior than the foliage tests. It is because they indicate the nutritional states and fertilizers needs of nursery soils. The other objectives of soil test area;

1. To maintain the fertility status of nursery beds; and
2. To predict the response of beds to the application of time and fertilizer.

The following nutrients are of great interest, viz.,

1. Cations (Ca, Mg, and K); and
2. P (Phosphorus)
3. Organic matter content; pH value, and the levels of salts in the soil.

A representative sample is the very critical part of the soil testing in the care of nursery. It is advised that all areas which vary with respect to the following must be sampled separately, viz.,

1. Appearance;
2. Slope;
3. Drainage;
4. Soil type; or
5. Past treatments, if any

The fertility test should be conducted/performed to assure that all the essential elements are present in sufficient quantities, examples are present in sufficient quantities, examples of soil fertility sandards for forest nurseries are given as under:

Soil property	Sands to loamy sands	Loamy sands to sandy loams	Silt loam to loams
Organic matter %	1.50	2.00	> 3.00
N %	0.07	0.10	0.15
pH	5.3-5.8	5.3-5.8	5.3-5.8
P kg/ha	55-110	85-110	85-140
K kg/ha	85-140	140-200	170-280
Ca kg/ha	450-670	670-1000	> 1,000
Mg kg/ha	55-65	65-100	>100

Source: Simpsom, J.A. 1978, Nursery Nutrition Studies with slash pine at Beerburrum queens land Res. Paper 9. Brisbane. Queensland, Australia, Department of Forestry. 32 p.

Fertilizers

Fertilizers are defined as, materials having definite chemical composition with a higher analytical value and capable of supplying plant nutrients in available forms. However, in generally notion the fertilizers are inorganic in nature and most of them are the products of different industries. Only the exceptions are inorganic nature.

Classification

The fertilizer are classified as follows:

(*i*) Inorganic or organic;

(*ii*) Natural or synthetic.

Davidson and Mecklenburg (1981) gave the chemical and physical properties of the fertilizers, viz.,

Chemical Properties	Physical Properties
A. State:	**A. State:**
1. Organic	1. Dry
(a) Natural	(a) Pulverized
(b) Synthetic	(b) Granular prills
2. Inorganic	(c) Tablet
(a) Natural	(d) Packet
(b) Synthetic	(e) Encapsulated
B. Element:	2. Liquid:
1. Single	(a) Water soluble
2. Double	(b) Suspension
3. Triple (Complete)	**B. Release Rate:**
4. Multi.	1. Rapid
C. Long-term effect on soil acidity	2. Slow (Controlled)
1. Acidic (Lower pH value)	
2. Natural	
3. Basic (Raises pH value)	

Other types

1. **Complete fertilizer:** Complete fertilizer is referred to a fertilizer material which contains all three major nutrients, N, P and K. These are expressed as percentages of elemental N (Nitrogen), P_2O_5 (Phosphate) and, K_2O (Potash). Therefore, a label with 5-10-5 means 5% N, 10% P_2O_5 and 5% K_2O. Further, ion 1 kg there is 50g of N, 100g of P_2O_5 and 50g of H_2O.
2. **Incomplete fertilizer:** This pertains to a fertilizer material which lacks any one of three major nutrients (N, P and K).
3. **Straight fertilizers:** May be defined as chemical fertilizers which contain only one primary or major elements, for example, Ammonium sulphate $(NH_4)_2SO_4$; Urea $[CO(NH_2)_2]$
4. **Mixed/complex fertilizer:** It may be defined as a fertilizer material which usually contains more than one primary/or major nutrients produced by the process or chemical reactions.

There are generally three types of chemical fertilizer available in the market namely,

1. Nitrogenous;
2. Phosphatic; and
3. Potassic fertilizers.

Nitrogenous fertilizers

Alternatives: There are three alternatives for applying the fertilizers, viz.,

1. Mix dry fertilizers directly with the growing medium. This alternative is commonly used for dry/commercial fertilizers applied before sowing or transplanting. Dry/commercial fertilizers can also be broadcast by hand.
2. Put soluble fertilizers solutions on the soil. This alternative is ideal for small nurseries requiring minimal amounts or for very large nurseries having good irrigation system.
3. Formulating nutrient solutions on sites. This alternative involves making homemade fertilizer solution to be applied by hand sprayers or injected into irrigation systems. Home-made solution are not practical for small nurseries.

Classification: The nitrogenous fertilizers can be classified into four classes on the basis of forms of N present in straight nitrogenous fertilizers as follows:

1. Nitrate nitrogen (NO_3-N) containing fertilizers, for example, $NaNO_3$ – 16% M'; and

   $Ca(NO_3)_2$ – 15.5% N.
2. Ammonium containing nitrogenous fertilizers (NH_4-N), for example, $(NH_4)_2SO_4$ – 20% N;

   NH_4Cl – 24 to 26% N; Anhydrous ammonia, 82% N.

3. Both NH_4 and NO_3 – N containing nitrogenous fertilizers, e.g., ammonium nitrate (NH_4NO_3) – 33 to 34% N; Calcium Ammonium Nitrate (CAN) – 20% N; and
4. Amide fertilizers-are the organic form of N-containing fertilizers, for example, urea [$CO(NH_2)_2$] – 46% N; Calcium Cynamide ($CaCM_2$) – 20% N.

Most common N-fertilizers: Following are the most common N-fertilizers which are usually used by nurserymen, viz.,

1. **Ammonium sulphate [$(NH_4)_2SO_4$]:** This contains about 20% N, and it is white crystalline salt, stable and soluble in water. Moreover, it can be stored very well and for longer periods. Its solution is acidic to litmus.

   Reactions in Soils: For example, when ammonium sulphate is applied to calcareous soils of nursery, considerable loss of ammonia through volatilization may occur:

$$(NH_4)_2SO_4 + CaCO_3 + 2H_2O \rightarrow$$
$$\downarrow CaSO_4.2H_2O + (NH_4)_2CO_3.$$
$$(NH_4)_2CO_3 ¾\!\rightarrow 2NH_3\uparrow + CO_2\uparrow + H_2O.$$

   The ammonium carbonate $(NH_4)_2CO_3$ is unstable and decomposes easily to go NH_3, CO_2 and H_2O. The loss of nitrogen from $(NH_4)_2SO_4$, applied to broad cost in soils may be worth mentioning.

2. **Urea [$CO(NH_2)_2$]**

   It is an organic solid nitrogenous fertilizer. It is manufactured by reacting anhydrous NH_3 and CO_2 gas under very high pressure in the presence of a suitable catalyst. The probable reaction is as follows:

$$2NH_3 + CO_2 \rightleftharpoons NH_2COONH_2$$

(Ammonium carbonate unstable products)

   This unstable intermediate product is decomposed and area is recovered. The curve solution is then concentrated to 99% and sprayed into a chamber where area crystals are formed.

$$NH_2COONH_4 \rightleftharpoons CO(XIH_2)_2 + H_2O$$

(Ammonium carbonate) (Urea)

   Reaction in soil: After application of urea to soil, it undergoes enzymatic hydrolysis to product an unstable compound designated as ammonium

$$NH_2 - CO - NH_2 + H_2O \rightarrow NH_2COONH_4$$

Ammonium carbonate

(Unstable)

$$NH_2COONH_4 \rightarrow NH_3 + CO_2$$

   This NH_3 is converted to NH_4^+ ions by accepting one portion (H^+) from proton donor and therefore, subsequently from NH_4OH.

Phosphalic fertilizers

The plant nutrient content of all phosphatic fertilizers is expressed as percentage of phosphorus rather than the percentage of P_2O_5. The conversion of P% P_2O_5 and vice versa (% P_2O_5 + to % P) is very simple and usually expression are used:

Classification: Phosphatic fertilizers can be classified into three groups on the basis of forms in which orthophosphoric acid acid (H_3PO_4) is combined with calcium (Ca^{2+}).

1. **Water soluble mono-calcium phosphate [$Ca(H_2PO_4)_2$], for example,**
   (a) Superphosphates (SSP, DSP, and TSP)
   (b) Single superphosphate (SSP): 16 – 18% P_2O_5 or 6.88 – 7.74% P;
   (c) Double superphosphate (DSP): 32% P_2O_5 or 13.76% P;
   (d) Triple superphosphate (TSP): 46 – 48% P_2O_5 or 19.78 – 20.64% P.
2. **Citric acid soluble, di-calcium phosphate [Ca_3 $2H_2(PO_4)_2$ or $CaHPO_4$], for example,**
   (a) Basic slag, silicats of times: 14 – 18% P_2O_5 or 6.02 – 7.74% P.
   (b) Di-calcium phosphate: 34 – 39% P_2O_5 or 14.62 – 16.77% P.
   (c) These are not soluble readily in water and not readily available to the plants.
3. **Phosphatic fertilizers not soluble in water or not soluble in citric acid, tricalcium phosphate [$Cs_3(Po_4)_2$].**
   For example,
   (a) Rock phosphate, 20-40% P_2O_5 or 8.6 – 17.28% P;
   (b) Rain bone meal, 20-25% P_2O_5 or 8.6 – 10.75% P;
   (c) Steamed Bone meal, 22% P_2PO_5 or 9 – 46% P.
   (d) Super phosphate is formed due to the reaction of rock phosphates and conc. H_2SO_4 as follows:

$$Ca_3(Po_4)_2 + 2H_2SO_4 + 5H_2O \rightarrow \underset{\text{(mono-calcium phosphate)}}{Ca(H_2Po_4)_2} + H_2O + CaSO_4.2H_2O$$

Potassic Fertilizers

Now-a-days, supply of potassium to the plants are usually expressed as K^+ (potassium) percentage instead of K_2O percentage in potassic fertilizers. Converting percentage K to percentage K_2O and vice-versa can be expressed as,

$$\%K = \% K_2O \times 0.83;$$
$$\%K_2O = \% K \times 1.2$$

Muriate of Potash (KCl)

The term muriate from muriatic acid a common name for hydrochloric acid. The fertilizer grade muriate contains 50–52% potassium (60-63 per cent K_2O). Muriate of potash is generally marketed in five particles sizes;

1. Special standard;
2. White soluble;
3. Standard;
4. Course; and
5. Granular.

Sulphate of potash (K_2SO_4)

It is a white salt which contains 41.5-44.2% K(50.0 –53.2% K_2O). It is produced by a number of processes, like,

1. **Langbeinite process:** The production form langbeinite minerals ($K_2SO_4.2MgSO_4$, 22.6% K_2O), occurs according to the following reaction:
   $K_2SO_4.2MgSO_4 + 4KCl \longrightarrow 3K_2SO_4 + 2MgCl_2$.
2. **Trona process:** Bar kite ($Na_2CO_3.2Na_2SO_4$) is reacted with KCl to form to form glaserite ($Na_2SO_4.3K_2SO_4$) and then reacted with KCl brine to give potassium sulphate (K_2SO_4).
3. **Galserite process:** The following reactions take place in the production of sulphate of potash from Na_2SO_4 (Sodium sulphate) and KCl (Potassium chloride) with glaserite as an intermediate product.
   $4Na_2SO_4 + 6KCl \longrightarrow Na_2SO_4.3K_2SO_4 + 6NaCl.$
   (glasrite)
   $Na_2SO_4.3K_2SO_4 + 2KCl$ ¾→ $4K_2SO_4 + 2NaCl$
   (glasrite)

Reactions in Soils

Both chloride and sulphate of K are soluble in water (H_2O) and application to the soil they ionize into; K^+, Cl^- and SO_4^{2-} ions. Therefore, the released K^+ ion from the fertilizers gets adsorbed on the soil colloids and also available to the plant through cation exchange reactions.

Complex Fertilizers

Complex fertilizers may be defined as a material containing all three primary nutrients, (viz., N, P, and K) and it is also designated as complete complex fertilizers, whereas a fertilizers material containing only two of the primary nutrients, say N and P or P and K, etc. and these are usually known as incomplete complex fertilizers.

The most important complex fertilizers are nitrophosphate, polyphosphate based fertilizers, most important grades of nitro phosphate is 20:20:0 and that means nitro phosphate contains 20% nitrogen, 20% phosphours and no potassium.

In addition to these, there are various other commercially produced complex fertilizers salphate of different grades, for examples,

15 : 15 : 15;

20 : 20 :20; and

10 : 26 : 26.

Polyphosphate fertilizers

1. Ammonium polyphosphate;
2. Potassium polyphosphate;
3. Zincated polyphosphate; etc.

Bio-Fertilizers

The bio-fertilizers are also called microbial inoculants.

Definition

These are usually defined as, "preparation containing live or latent. Cells of efficient strains of nitrogen fixing, phosphate, solubilizing or cellulolytic micro-organisms used for application to seed, soil or composting areas with the objectives of increasing the population of such beneficial micro-organisms and accelerate certain microbial processes to augment the extent of the availability of nutrients in a form which can be easily assimilated by the plants.

There are several types of bio-fertilizers, viz.,

1. Rhizobium - for leguminous crops;
2. Azotobacter- for non-leguminous crops;
3. Azospirillum - for non-leguminous crops;
4. Blue-green bacteria (Cyanobacteria);

Azolla (water forms, forms a green mat over water which quite often become reddish due to the accumulation of anothocyanion pigments). These car fix nitrogen and acts as a bio-fertilizers augmenting nitrogen in the soil.

Further,

1. The fertilizers preparation with rhizobium culture is popularly known as 'nitorgin' and;
2. The bacterial fertilizer of azatobacter chroocoum is, however, known as 'Azatobacterin'.

The most prominent and popular rhizobial cultures (Rhizobium japanicum, R. mililotic, etc.) must have the following specifications:

1. The inoculants shall be carrier based;
2. The culture must contain a minimum of 10^8 viable cells of Rhizobuim per gram of the carrier on dry mass basis within 15 days of manufacture and 10^7 within 15 days before the expiry date (say maximum of six months);
3. The pH of the culture shall be between 6.0 and 7.5; and
4. The carrier material should be in the form of powder, for example peat, lignite, peat soil, humus, etc.

Now, the used of Rhizobial cultures in t he establishment of legumes has become popular and widely accepted.

Azolla can also be generally used as a bio-fertilizer and some other most important species are, *Azolla Carolinian*, *A. nolotica*, *A. filiculodies*, *A. mexicama*, *A. microphylla* and *A. pinnata* of which A. pinnate is the most common in India, *Azolla inoculums* @ 0.1 to 0.4 kg per sq.m is most ideal for rapid multiplication of the water form in the nurseries for production of 8-10 t/ha green matter in just 20 days. However, usually two methods of Azolla application have been recommended in India;

(a) as a green manure applied in filedsd before transplanting of rice; and

(b) by dual cropping with rice when azolla growsn side by side with rice for some time. It has been noticed that the application @ 10 t/ha fresh type of Azolla is as efficient as basal does of 25-30 kg N/ha. These application may also be tired in the nursiers of forest crop for getting very promising results.

Chapter

11

MAINTAINING SOIL FERTILITY

INTRODUCTION

The plants are raised in the nursery year after year and by doing so that fertility of the soil dwindles down and ultimately this has a great adverse effect on the quality of the plants/seedlings. Therefore, the fertility of the soil of a nursery is very essential to maintain at an optimum level. The following are some of the methods and ways to adopt for the purpose:

1. **To keep the land fallow:** It is a good exercise just to leave the land fallow for some time and it will help in restoring the fertility of the nursery soil. Therefore, if, 1/4 (one-fourth) of the area of nursery may be left follow by rotation for a year, then all parts of the nursery would get rest on a 4 year cycle and this would help them regaining their original fertility to some satisfactory extent.
2. **Use of well-rotten leaf mould:** By mixing of well-rotten leaf mould in the soil may help to restore the fertility. The leaf mould in the form of well-rotten condition, is usually easily available in temperature forests whereas in tropical forests, it may prepared by putting the dry leaves in the pit. This type of pit is covered with soil. With the passage of time about few months it will not and thereafter it can be easily used as manure which is essential for the growth of the seedlings and plants.
3. **Use of green manuring:** Green manuring is nothing but can be defined as, turning a crop under, while green or immediately after maturity, for the purpose of soil improvement.

   The leguminous crop is turned by sometimes the non-leguminous crop is also used. In Indian conditions, the following crops are used as green manuring in forest nurseries:

   (a) *Crotolaria juncea*;

   (b) *Sesbaria aculeate*; and

   (c) *Cannabis sativa*.

   However, green manuring is usually not recommended for the following two conditions, viz.,

   (a) Regions of low-rainfall; and

   (b) Where irrigation is not available.

   Moreover, mixing of well-rotten farm yard manure (F.Y.M>) also helps the soil to regain its original fertility.

4. **By mixing compost:** Compost is defined as, a mixture of vegetable materials, which has been allowed to rot, often with the addition of animal and/or mineral products. It is used as a soil improver. The compost is made by rotting the vegetable wastes usually in a pit of 3m × 2m × 1m but the size of pit may vary according to the requirements.

   However, alternatively, well-rotten manure can also be added to accelerate decomposition. The pit may be opened in about four months. The compost contains about 3.2% nitrogen and 6.0% phosphoric acid. The organic manures have to be used at the rate of 125 tonnes per hectare.

5. **By using chemical fertilizers:** It is not so easy to arrange for optimum quantities or organic manures, the chemical fertilizers re increasingly used. Before using fertilizers, nutrient starts of the soil should be assessed. The test should not only relate to major nutrients but also for trace elements.

SOIL HEALTH

The soil health is also called as soil quality. The soil quality is defined as, the capacity of a soil to function, within ecosystem and land use boundaries to sustain productivity, maintain environmental quality, and promote plant and animal health".

Soil Functions

The important soil functions related to crop production usually include:

1. Infiltration/storage of water;
2. Retention and cycling of nutrients;
3. Weed suppression;
4. Pest control;
5. Detoxification of harmful chemicals;
6. Sequestering of carbon; and
7. Production of food and fibers.

Characteristics of Healthy Soil

The healthy soil has some good characteristics which have great importance in keeping the status of soil fertility. Following are some characteristics of healthy soil:

1. Good soil tilth;
2. Sufficient depth of soil;
3. Sufficient but not excess supply of nutrients;
4. Small population of plant pathogens and insect-pests;
5. Low need infestation pressure;
6. Good soil infestation pressure;
7. Large population of beneficial organisms;

8. Resistant to degradation;
9. Free of chemicals and toxins that may harm the crop; and
10. Resilience when confavourable conditions

Soil Quality Index

Today the concept of soil quality has been just broadened to include attributes for food, safety and quality, human and animal, health and environmental quality soil quality index is a function of the following parameters:

Soil Quality Index (SQI)

= f(SP. P. E. H. HR. BD. FD. MI)

Where, SP = Soil properties;

P = Potential productivity;

E = Environment factors;

H = Health of animal and human

ER = Erodibility;

BD = Biological diversity;

FQ = Food quality

MI = Management impacts

The primary purpose of forest nurseries is to produce trees to form new forests. Hence, maintaining adequate fertility in the nursery soils is important to assure production of high-quality planting stok.

Soil pH

Soil pH or reaction is described by the pH scale in which 7.0 is neutral for soils measured in water.

1. pH 5.5 to 6.5 — Slightly acidic;
2. pH 4.5 to 5.5 — Acidic;
3. pH < 4.5 — Strongly acidic
4. pH < 3.5 — Rare soils
5. pH > 7.5 — Alkaline soils

The soil pH usually the availability of nutrients to plant and influences the composition of soil flora and fauna including some crop pathogens. The ideal values for coniferous soils are pH 5.0 to 6.0 and therc of broad leaved nurseries soil pH 6.0 to 7.0.

Nutrient Form

1. **Nitrogen:** Three forms of N occurs in soil:
   (a) Organic N is associated with the soil humus;

(b) Ammonia N(NH_4 – N) fixed within the lattice of clays such as vermiculite;

(c) Soluble inorganic ammonium and nitrate compounds.

The soil organisms slowly down organic matter, ammonium is released into the soil solution. However nitrification is slow in forest nursery soils because most but not all are lów in bases. However, most conifer seedlings grow very well with a predominantly NH_4–N source.

2. **Phosphorus:** This is usually present in the soil mainly as inorganic phosphates (50 to 70%) and organic P(30 to 50%) which together compose the solid phase. These soluble phosphates react with Fe and aluminum (Al) and acid soil conditions to form insoluble Fe PO_4 + $2H_2O$ and $AlPO_4$ + $2H_2O$ and with Ca to form apatites in neutral sols and also calcareous soils. The P, thus becomes fixed in a form that is unavailable to the plants. Phosphorus fertilizers dissolve in water to release orthophosphate. Organic matter improves P availability because it;

   (a) Competes with phosphate ions for binding sites on soil particles; and

   (b) Produces organic anions which chelate Al, Fe and Ca and;

   (c) Slowly releases P during decomposition.

3. **Sulphur** (S) occurs mainly in soil organic matter and is absorbed by plants as sulphate. The amount of sulphur in a conifer is closely related to the amount of N; the ratio is 1 (one) part 5 to 14 parts N by weight;

   The cations K^+, Ca^{++} and Mg^{++} are adsorbed on soil cation-exchange sites where they are available to the plants. However, even K(Potassium) can be rapidly depleted by leaching in sandy soils nursery with a pH of 5.0 or less.

FERTILITY MANAGEMENT

1. **Controlling soil pH:** It has been noticed that raising soil pH may be somewhat easy whereas to reduce it, is much more cumbersome task. Therefore, for increasing the pH, one must be careful and conservative. However, the fertilizers can modify soil pH of nursery soil with addition of nitrogen (N) and P(Phosphorus) sources leading to acidity. Further, soil pH may be reduced with S(Sulphur), Al (aluminum) sulphate [$Al_2(SO_4)_2$], and sulphuric acid (H_2SO_4). But these substance prove to be somewhat toxic to the seedlings. Ground timestone or dolomitic limestone (which contains Mg^{++} as Ca^{++}) is equally good raising pH of soils of nurseries.

2. **Organic matter:** Davey and Krause (1980) advocated that organic matter usually consists of three major components, viz.,

   (a) Plant, animal and microbial residues in various stages of decomposition;

   (b) Humus; and

   (c) Live microorganisms.

   The organic materials, when added to the soils of nursery, are usually decomposed by microorganisms which respire as their respiration process causes a large part of the organic matter to be lost by oxidation. Various forms of organic amendments are added to soils of nursery but the addition C(carbon) can easily increase the C;N ratio to reduce the amount of N(Nitrogen) which is available to the crop of forest. Dry softwood saw dust can immobilize about 6kg of nitrogen (N) per tonne (Allison 1965). Therefore, supplemental N fertilization is necessary when some organic amendments are made in the nursery soils. Rogers and Giddens (1957); Davey and Krause (1980) reported that cover-cropping that never or the soil nitrogen (N), may however, benefit the soil by just conserving nutrients otherwise are lost by leaching and improvising the soil physical and biological properties.

3. **Fertilization:** The fertilizers can be,
   (a) Organic: Such as compost or manure;
   (b) Inorganic: The various salts of nutrients elements now widely used in forest nurseries.

   The counteractions of nutrients elements in organic fertilizers are generally found as low, for example, compost may contain 2-4% nitrogen and 0.2 – 1.8% phosphorus (Benzian 1964) and farm yard manure about 1.1 – 1.5% nitrogen (N). The soil pH can be changed by just adding mineral fertilizers. The ammonium and urea salts make the soil more acidic whereas the nitrate fertilizers which contain a base of KNO_3 or $Ca(NMO_3)_2$ can increase the soil pH of nurseries of forests. It has been observed that potassium sulphate and chloride have no effect on soil pH.

NUTRIENT ELEMENTS

The application of fertilizers of soils of nurseries is very important and essential for replacing the lost nutrients from the soils due to some or the other regions weeding operations performed by hand also cause a huge loss of nutrients from the soils. In actual practice, only the seeds and part of the roots are removed and the reaming things are left there itself just for getting decomposed in the soils which may serve as fertilization in the form of nutrients. In conifer crops/seedlings, the nutrients removed may vary from,

Nitrogen (N) = 50 to 200 kg;

Phosphorus (P) = 4 to 35 kg and

Potash (K) = 25 to 105 kg

1. **Nitrogen:** It is quite evident that most of the conifers respond very quickly to nitrogen (N) fertilizer. However, in general, the earlier N(Nitrogen) is applied, the better it may prove. The early may sown seedlings, generally benefit from a top dressing nitrogen fertilizer in late-June. However, where there is a problem of damping-off in the soils of nursery bed, application of nitrogenous fertilizer is usually discarded. However, many applications of 22 to 44 kg/ha of nitrogen (N) should be applied during second growing season as a top dressing. The seedlings/nursery beds should be watered soon after dry soluble nitrogen fertilizer was applied.

2. **Phosphorus:** It is also very essential element for the good growth of seedlings. Therefore growth responses to phosphorus fertilizers are seldom defected, if, maintenance dressing of 67–135 kg/ha, of phosphorus are applied each rotation. The phosphorus fertilizers should be applied to the soil very close to seedlings roots. However, top dressing of phosphorus are somewhat ineffective except on very sandy soils.

   Calcium superphosphate (0-20-0) is a very promising fertilizer for acidic sandy soils because it supplies calcium and sulphur and also phosphorus and may reduce the soil acidity. Ammonium phosphate, with grade 11-55-0, must be used, if the superphosphate does not provide sufficient phosphorus nutrition.
3. **Potassium:** The positive growth responses of conifer seedlings to potassium fertilizers re rarely detected. However, both quantity and frequency of application summed to affect the growth of seedlings. The frequent top dressing of potassium fertilizer throughout the second growing season was supposed to solve the problem. Van dem Dressche (1963) reported that excessive potassium fertilization can results in undesirably very high soil potassium levels that may reduce the growth of seedlings.

Application Methods

1. The fertilizers may be applied in nurseries in a number of ways. It depends upon, (*i*) Time of treatment; and (*ii*) Nutrients which are supposed to be applied. For instance, nitrogen is generally required in huge quantities, therefore, must be top dressed, whereas phosphorus should be placed very near to the root and must be applied before sowing the seeds in the nursery beds.
2. The fertilizers may be broadcast with some types of agricultural spreaders, when phosphorus or calcium fertilizers are applied seed shaping with rotary fingers. A good watering is essential after top dressing to just wash fertilizer-off the seedlings for preventing the burn which is caused due to the doses of fertilizers.
3. Fertilizers can also be applied to foliage through over-head irrigation systems towards the end of irrigation time. In this way, the distribution of irrigation water may prove to be uneven, CFASIC (1980) advocated that pesticide spray equipment of high-pressure, low-volume is found very satisfactory for failure feeding.

Nitrogen in the form of urea was found very common faliar applied nutrients. This method gives quick responses. Urea is usually applied in solutions containing 400-800g/100 liters. It is advised that applications should not made in the add weather. However, only 5% concentration (Weight/Volume, i.e., 5kg/100 liters) is generally advised.

Responses of Seedlings to Nutrients

The responses of seedlings to nutrients is described as follows:

1. **Growth:** Etter (1971) suggested and advocated that newly germinated seedlings of conifers may here sufficient nutrients and exhibit little (very small) response to different levels of nutrient supplied from outside sources for upto nearly six weeks after germination. The nutrients are needed regularly in the whole period of the season, the rate of phosphorus uptake, varies to a great extent in some conifers. However, by contrast, nitrogen and potash uptake rates are highest at the beginning of the growing period of season and thereafter if decreases slowly. It has been seen and noticed at several occasions, that with holding of nitrogen (N) or phosphorus (P) goes to restrict shoot growth more than the root systems. Therefore, additional fertilization is sometimes necessary to offset the effects of wet or cold weather. In such cases, the fertilizers are leached is reduced by the effect of low temperature.
2. **Drought stress:** The drought conditions are very crucial in nurseries of forest. They may have detrimental effect on the growth and survival of the seedlings. Pharis and Kramer (1964) observed that intermediate nitrogen levels have either been beneficial or have no effect whereas low nitrogen levels have had the least effect on tree growth. Further, high levels of N (nitrogen) levels are supposed to promote shoot growth of seedlings. The seedlings with large shoots may transpire more water than those of small shoots. Moreover, high levels of foliar potassium are concerned with the reduction in transpiration rates in the trees. Increased concentration of K (Potassium) showed an increase in the drought survival of prince.
3. **Cold hardiness:** The nutrient may influence the seasonal growth patterns of seedlings. This may, however, alter the seedling susceptibility to the low temperature conditions. A heavy dose of nitrogen fertilizers is usually found to delay dormancy. Further, the time of fertilizer application influences the outcome. Timmins (1974) found that after a hardening period young seedlings with an internal K:N ratio (potassium : nitrogen) of about 0.6 were somewhat hardier than those with a ratio of 1.3.

NURSERY REGISTER

The nursery register is a very essential document to be maintained upto-date. Therefore, every nursery should have a nursery register giving full information regarding the nursery. In the very beginning, this register should impart the general particulars about the nursery, viz.,

1. Name of the nursery;
2. Location of the nursery;
3. Year of establishment;
4. Area of nursery (in hectares/acres etc.);

5. Soils (its condition, type, class, etc.);
6. Method of irrigation;
7. Type of fencing (with initial cost);
8. Manuring (its details and cost);
9. Soil working done at the time of initial formation with cost; and;
10. Rent, if any.

This description must be supported by a good map a nursery showing the following details.

1. Lay-out;
2. Allocation of portion to species; and
3. Plot and nursery bed serial numbers.

Some additional information should also be given in the nursery register, viz.,

1. Quantity/number of seeds sown in a bed;
2. Date of seed sowing in the nursery;
3. Date of germination of seeds;
4. Weedings with dates;
5. Irrigation with dates;
6. Manuring/fertilization, etc.

Khanna (1984) mentioned somewhere in forestry literature that the nursery register should also give the details of all expenditure incurred so that the cost per plant could be estimated and calculated. Further, this should also give the results of counting nursery stock in each and every nursery bed on,

1. Ist of January; and
2. Ist of June every year.

Therefore, at the time of counting, the planets should be usually classified in two categories, viz;

The first are for use in the coming planting season, moreover, it should also give information about the plants form the nursery to planting out site/area in the coming planting season of the year. Further, the balance of stock of seedling should also be mentioned in the nursery register. The nursery men/nursery managers must be very vigile in keeping all the information in the nursery register which is a good document for further use.

GLOSSARY

Abnormal seedlings: In seed testing, seedlings which donor possess al normal structures needed for growth, nor show the capacity for continued development.

Abscission; Natural separation of leaves, flowers, and fruits and fruits from plant, generally associated with deterioration of a specialized layer of thin walled cells.

Absorption: Movement of a pesticide from the surface into a body of water or of nutrients into a plant.

Acid Soil: Soil having a pH value less than 7.0.

Adjusted sodium adsorption ratio: Also, called ASAR. Irrigation-water quality index that is the ratio of deleterious ions (i.e., Na^+, Co_3^- and HCO_3^-) to beneficial ions (i.e., Ca++ and Mg++). The ASAR value is valued to asses the effect of salts on solid permeability.

Aeration (Soil): Process by which (1) oxygen diffuses through the soil to the root, and (2) carbon dioxide and other gases released from the root diffuse to the soil surface.

Aerobic: Growing or occurring in the presence oxygen.

Alkaline soil: The soil having a pH value greater than 7.0.

Anaerobic: Growing/occurring in the absence of oxygen.

Anion: The ion having a negative charge, e.g., Cl^-, No_3^-.

Annual: The plant which completes its whole site cycle from the seed germination to seed production and death within a single season.

Artificial Regeneration: Reforestation of a stand by direct seeding or planting.

Assimilation: Portion of the soil water which can readily be absorbed by the plant roots, generally considered to be that water held in the soil within water-potential range of a approximately –1/3 to –15 bars.

BAR: Unit of pressure equaling 10,00,000 dynes/cm^2; tropically used to measure plant and soil water potential.

BED: Elongated strip of soil in which seedlings or transplants are usually grown.

BED Foot (m): Area of seed bed/lined foot (or 1 lineal metre) longtimes the width of the bed.

Biennial: Plant that requires 2 years to complete its entire/whole life cycle; from seed germination to seed production and death. The food is stored in the first season and then used in the second season when flowers and seeds are produced.

Broad Cast Application: spreading of a chemical or the fertilizer over the whole area or field rather than only on rows.

Broadcast Seeding: Method of sowing in which seeds are distributed across the seedbed by hand.

Broad Leaf Species: These plants classified as dicotyledoneae; characterized by having round or flattened leaves.

Buffer Capacity: Ability to resist change in pH. A soil with a high buffer capacity will have stable soil pH.

Buck Density (Soil): Mass/weight of dry soil divided by soil volume, commonly expressed as grammes per cubic centimeter

Caliper: Stem Diameter of a seedling usually measured just above the root collar.

Capillary Water: Water held in the small pores of a soil usually with attraction forces exceeding the pull of a 60 cm column of water. However, most capillary water is available to the plants.

Carabiner: A metal safety clip, used by climbers, with repes, which can be locked in the closed position as an insurance against accidental opening during climbing and fruit/seed harvesting.

Carbon Nitrogen (C.N) Ratio: Ratio of carbon content to nitrogen content. The C.N. Ratio in plant residue is often a convenient predictor of decomposition rates but is not the only determinant.

Catch Crops: The crops which are grown and then incorporated into the soil, primarily to capture and retain nutrients on the site but also to increase soil organic matter.

Cations: The ion having a positive charge, e.g., Ca^{++}, Mg^{++}.

Cation Exchange Capacity (CEC): Sum total of exchangeable cations that a soil can absorbs, expressed in milke equivalents per 100 grammes of soil clay, or organic colloid. Because of the high CEC of organic matter, the buffer capacity of a soil is increased after organic amendments are added and the soil is just able to hold cationic nutrients.

Cotyledon: Modified leaf or leaves of the embryo or seedling, which may contain the stored food reserves of the seed. They, however, are formed at the first node or at the upper end of the hypocotyls.

Cover Crops: Crops grown principally to control various forms of erosion but also incorporated into the soil to increase organic matter.

Cull: The seedling which is not acceptable because it does not meet some size and quality standards and which is through to have low survival and growth potential.

Cull Factor: Number of seedlings that donot meet shippable, e.g., diseased, poor form, or size, damaged, expressed usually as a percentage.

Cuctural Control: Indirect control measures, as opposed to direct killing, used to prevent pest damage-e.g., use of a previous crop that is antagonistic to the pest.

Damping-off: A disease characterized by either seed decay in the soil or seedling wilting and death after germination, usually caused by soil-borne fungi.

Decdious Plants: The plants which seed or leaves at a certain season, generally autumn.

Designed Experiment: A detailed investigation in which the experimenter or researcher requires precise and unbiased (1) conclusions, and (2) measure of uncertainty associated with those conclusions. The treatments are usually replicated and applied to plots at random.

Desorption: Loss of moisture from a relatively moist-hydroscopic material, such as seeds, to a relatively dry atmosphere until to two reach equilibrium.

Determinant Growth: A two-year process of shoot growth in which primordial are laid down in buds in the first year but elongate in the second year. The species exhibiting determinant growth may only have only one growth flush per year.

Dicot: Larger of the two classes of flowering plants (Angiospermae), distinguished from the smaller class (monocotyledoneae) by the presence of two seed leaves (cotyledons) in the embryo and by other structural features, e.g., met-veined leaves, includes oaks, elms alder and some herbicides, e.g., 2,4-D, are effective against dicots but usually harm monocots.

Dormancy: A physiological state in which a seed predisposed to germinate does not, even in the presence of favourable environmental conditions.

Dormancy Combined: Dormancy as result of two primary factors, such as seed coat conditions: impermeability to gases or moisture or mechanical restrictions.

Drench: Saturation of a soil with pesticides generally to control root diseases.

Drill Seeding: Nursery sowing methods in which seeds are planted in the rows with a seed drilling implement.

Effective Kilogram: The weight of seed of any particular seed lot which can be expected to produce the same number of either viable seeds or plan table plants as would be producers by one kilogram of standard seeds.

Empty Seed: A seed without any content, or without an embryo or embryo cavity, if, some residual tissue is present; the opposite of filled seeds.

Enzyme: A catalyst produced by living organisms and acting on one or more specific substances.

Epigeal Germination: Germination in which the cotyledons are forced above the ground by the elongation of the hypocotyls.

Exocalp: Outermost layer of pericarp; the skin on fleshy fruits as in cornus, malus, and prunus.

Exogenous: Developed from a superficial layer of tissue. In seeds, developed from or occurring in the covering of the seeds, either seed coat or pericarp.

Fallow: Allow cultivated land to lie idle during the whole or greater portion of the growing season.

Fermentation: The process of chemical changes in organic substances caused by the catalytic action of a 'ferment', which may be an independent plant such as yeast or bacteria, or an enzyme. May be accompanied by the production of heat and of toxic substances, hence the fermentation of fleshy fruits may adversely affect the seeds which they contain.

Fertilizer: Any organic or inorganic substance either of natural or synthetic origin which is added to the soil to provide elements needed for plant growth.

Field Capacity: Soil water content resulting after the free water has been allowed to drain from a saturated soil for one (1) or two (2) days; usually expressed as a percentage in a dry-weight basis.

Field Seed: A seed with all tissue essential for germination. Syn. Full seed.

Free Radial: A group of atoms which usually exists in combination with other atoms, but which may exist independently for shorter or longer provides, e.g., the NH_4 radical.

Frost Hardiness: Ability of plant tissues to survive and resist the stress from free sing temperatures without sustaining irreversible physical drainage.

Frost Heaving: Lifting of soil surface due to growth of ice-crystals in the underlying soils when this recurs over a period of time; seedlings can be physically lifted out of the ground.

Forst Pocket: Area whose topographic features cause cold air to accumulate increasing frost hazard to the seedlings.

Frost Smothering: Condition occurring when saturated soil freezes allowing little or no oxygen to reach plant roots. However, if, the conditions persist over 48-hours, seedlings damage or death can result.

Fumigant: Chemical applied as liquid or powder which volatilizes to gases and kills insects, nematodes, fungi, bacteria, seeds, roots, rhizomes, or entire plants, fumigants are usually applied beneath a tarp, sheet or other enclosures.

Fumigation: Use of chemical in gaseous form to destroy pests usually applied under a cover or shelter.

Fungiment: Chemical used to kill or inhibit the fungi.

Fungistasis: Inhibition of fungal growth, without destroying the fungus by preventing the germination of conidia and some other spore types.

Genotype: (1) An individual's hereditary constitution, with or without phenotypic expression of the one or more characters it underlies. The genotype is determined chiefly from performance of progeny and other relatives. It however, interacts with the environment to produce the phenotypes. (2) Individual characterized by a certain genic constitution.

Germination: In general, resumption of active growth in the embryo of a seed as demonstrated by the protrusion of the radical. In seed testing, (ISTA definition) resumption of active growth in an embryo which results in its emergence from the seed and development of those structures essential to normal plant development.

Germination Capacity: The proportion of a seed sample which might have germinated normally in a specified test period, generally expressed as a percentage. Syn., germination percentage. Somewhere, the term 'hyrmination capacity' has been used to just express the total of the seeds which germinate plus the ingeminated but sound seeds, as a percentage of the seeds sown, but this meaning is exclude by the latest INFRO definition (Bonner *et. al.* 1984a). Cf. viability per cent.

Germination Per Cent: Percentage of seeds that germinate under standard treatment and after a given tie period. This value is considered a principal index of seed quality is used to calculate seedbed sowing density.

Grey: Layer of mineral soil developed under condition of poor drainage, characterized by reduction of iron and other elements and gray colours and mottles.

Grading: A process of identifying and subsequently separating various classes of acceptable (shippable) and inferior (cull) stock to improve stock quality. This operation occurs after lifting and before packing and storing processes.

Green Manure Crops: Crop grown primarily as organic amendments for the soil, Green manure crop are incorporated into the soil while it is green but before seed set; to benefit success ding crops.

Gymnosperms: The botanical name of the group of vascular flowering plants which produce seeds not unclosed in an ovary, the most important order of which is the coniferales.

Hardening-Off: Natural process of adaptation by the plants to cold or drought. Hardening-off may be induced in the nursery by reducing water or by root-culturing, thus preparing the seedlings for cot-planting or transplanting.

Hard Pan: Hardened soil layer caused by cementation of soil particles with materials such as silica sesquixodies or calcium carbonate. The hardness does not change appreciably with changes in moisture content.

Hard Seeds: Seeds which remain hand and ingeminated at the end of a prescribed fest, because they have not absorbed water due to an impermeable seed coat.

Haustoria: Suckers of parasitic plants.

Herbicide: Chemical used to kill or inhibit unwanted plants or the weeds.

Hilum: Scar on a seed marking the point of attachment to the ovary in angiosperms or to the megasporophyll of gymnosperms.

Hormone: Growth regulating substance synthesized in one location generally in small amounts and then trans ported to another location within the organisms where it affects growth and differentiation.

Humic Acid: Mixture of dark coloured organic materials of indefinite composition extracted from soil with dilute alkali and precipitated by acidification.

Humification: Break down of organic residues to the humus.

Humus: Fraction of soil organic matter remaining after most animals and plants residues have decomposed; usually dark coloured; and the chemical composition of humus is very different from that of the original parent compound.

Humic substance: (1) help in soil retain water; (2) increase the cation exhcnage capacity (CEC), and (3) stabilize soil pH.

Husk: Outside envelope of a fruit; especially, if, coarse, harsh or rough as in the involucres of carya.

Hydroponics: Commercial production of plants in sand or gravel cultures. The sand or gravel is relatively inert, providing mechanical support for growth, and the nutrients are supplied by liquid solutions.

Hypocotyl: That of the axis of a germinating embryo which is between the cotyledons and the radical, in seedlings, the juvenile stem which is between the cotyledons and the rot system.

Hypogeal Germination: Germination in which the cotyledons remain in the seed below the ground while in the epicotyls elongates.

Imbibition: The mechanism of initial water uptake by the seeds; the taking up of fluid by a colloidal system.

Incorpation: Making of a fertilizer or some chemical into nursery soil before the operation of seed sowing.

Indeterminant Growth: The shoot growth resulting from successive periods of initiation and elongation of apical men stem cells during the growing season without extended periods of rest.

Infiltration Rate: The rate at which can be absorbed into a soil surface. However, it can be altered by nursery practices influencing the porosity and structure of surface soils. A soil with a poor infiltration rate is subject to surface run-off and erosion.

Inoculation: The process of introducing micro-organisms for some beneficial effect such as the addition of Rhizobium bacteria to legume seeds or of mycorrhiza fungi to nursery seedbeds.

Inoculum: Portions of pathogens, for example, fungal spores, capable of causing infection or initiating mycorrhizae upon contact with the host.

Interception System: Type of closed drainage system located below the water-table level, intended to remove water before it lenters the soil. The system may vary from a simple line of pipes positioned at the source of seepage to a complex grid pattern and is usually used in small land areas.

Ions: Atoms or groups of atoms which are usually electrically charged, that is, cations or anions.

Involucre: One or more whorls of bracts situated below and close to a flower or flower or flower cluster; sometimes enclosing the carpels as in Jectona, Castanea and Fagus.

Kernel: The nucellus of an ovule or of a seed, that is the whole body within the seed coats.

Kilogram effective: It is generally within as KEF. The ratio of seedling recovery from the standard seed to actual seedling recovery from a given seed lot.

Lateral Purning: Root culturing technique in which blades or colters are passed between drill rows to sever long lateral roots. The main purpose of lateral pruning is to facilitate lifting, stimulate root-growth and fibrosity and retard the height growth.

Leaching: It is the downward movement of materials in the soil solution. The soluble nutrients such as nitrate are often leached out of the seedling root zone.

Lifting Window: Time period of the year believed to be the best for harvesting seedlings from the seedbed, i.e., when the seedlings are most resistant to handle stresses and when subsequent survival and growth potential upon out-planting are high. The lifting window will vary from year to year depending on seed source, nursery location and cultural regimes used before lifting.

Lime (Calcium) Requirement: Amount of agricultural time stone required per area to raise the soil pH an optimal value of seedling growth; usually calculated for a soil depth of 15 cm or per 910000 kg of soil.

Liming: Addition of calcium sometimes including magnesium (dolomites) in the form of calcium carbonate; ground limestone or hydrated lime to furnish elements for plant growth and to neutralize soil acidity.

Loam: Textural class for a soil having moderate amounts of all three soil separates sand, self and clay.

Luxury Consumption: Absorption of excess amounts of nutrients these necessary for optimal growth.

Macerate: To soften by stepping in a liquid, with or without heat; to wear a way or separate the soft parts, by steeping.

Macronutrient: Chemical element necessary for plant growth in large amounts usually greater than 1 (one) part/500 in the plant; often supplied artificially in fertilizers.

Maintenance Dressing: Fertilizer application which functions to keep soil macronutrient and micronutrient levels adequate during the crop rotation.

Material Attributes: Individual measurements of one aspect of seedling quality which can indicate physiological conditions, e.g., leaf osmotic potential root starch concentration and shoot: root ratio. Performance attributes reflect the sum total of material attributes.

Matric Potential: Largest component of total soil-water potential caused by capillary action and attraction of water by soil particles. Metric potential can be measured with tension meters and is generally expressed in the negative pressure units such as bars.

Measurement Plot: That part of the experimental plot in which the observational units are measured.

Micronutrient: Chemical element necessary for plant growth in very small amounts. Micro-nutrient fertilizers are not normally needed except under soil conditions such as excessively high or low pH.

Mineralization: Conversion of organic elements to the inorganic state as a result of microbial decomposition.

Mineral Soil: Soil consisting largely of mineral matter, with organic matter generally less than 20%.

Moisture Content: The amount of water present in a material, e.g., wood, soils, or the seeds. It may be expressed in terms of weight of moisture as a percentage of the materials own-dry weight, ("dry-weight basis") or preferably in case of wet-weight including water (wet-weight) or fresh-weight basis.

Monocot: Smaller of the two classes of flowering plants (Angiospermae), distinguished from the larger class (Dicotyledoneae) by the presence of a single leaf (cotyledon) in the embryo and by other structural features, e.g., parallel-viewed leaves includes grasses, lilies, orchids. Certain herbicides are effective against monocots, but donot harm dicots.

Munch: Layer of plant residues or other material (e.g.k, plastic film, paper fiber) spread upon the soil surface to protect soils seeds, or plant roots from the effects of freezing, evaporation, crusting, etc.

Mycorhiza (E): The biological association, usually symbiotic between plant roots and particular fungi.

Naked Stratification: Pre-chilling of seeds without the use of a moisture-holding medium.

Narrow Leaf Species: Those plants classified as monocotyledoneae characterized by having narrow parallel veined leaves.

Natural Regeneration: Reforestation of a stand by natural seeding.

Nitrifaction: Biological process in which, (1) Ammonium is oxidized to nitrites, and (2) Nitrites are further oxidized to nitrates.

Non-Selective Pesticides: Material which is toxic to a wide range of pests or to more than one plant or animal.

Nucleus: A body of specialized protoplasm found in nearly all cells and containing the chromosomes.

Nut: Dry, indehiscent, one-seedbed fruit with a wordy or leathery pericarp developing from an inferior compound ovary.

Observational Limit: Item to be measure within are experimental plot, e.g., three seedling.

Operational Trial: Prelimary inquiry in which each treatment is applied to only one plot that is the treatments are un-replicated.

Organic Matter: The complex interaction of, (1) plant, animal, and microbial residues in various stages of decay; (2) humus; and (3) live organisms. Organic matter increases the buffer capacity, cation exchange capacity and water retention of the soil and provides a substrate for microbial activity.

Organic Soil: The soils usually containing 20% or more organic matter.

Outplanting: Planting of seedlings on a forest site.

Oxidation: Chemical process of combining with the oxygen removal of hydrogen or electrons.

Parasite: The organisms that live on some other living organisms.

Peak Germination: The point in time when rute of germination is highest; it may be derived in many ways.

Peat: Refers to largely un-decomposed or slightly decomposed organic matter accumulated under the conditions of excessive moisture and low oxygen availability; soil amendment used to increase soil organic matter and lower pH of the soil.

Percolation Rate: Downward movement of water through the soil, particularly the downward water flow in saturated or so nearly saturated soil. Percolation rate is used to calculate the internal drainage requirements of a soil.

Perennial: The plant which continues growing from year to year. Tops may die back in winter but roots or rhizomes persist.

Performance Attributes: The attributes of seedling quality measured by assessing the performance of seedlings subjected to environmentally controlled test conditions, e.g., root growth potential and frost hardiness.

Periodicity: The tendency, in an individual, stand or species, to produce seed at more or less regular intervals of more than one year.

Permanent Wilting Percentage: Water constant of a soil when the indicator plants growing tin that soil wilt and fail to recover when placed in a humid chamber varies with plant species but is generally considered to occur at approximately –15 bars water potential.

Permeability: The soil attributes that enable water or air to move through it determined by soil porosity.

pH: Numerical measure, negative logarithm of the hydrogen ion activity, of the acidity or alkalinity in a soil/solution. A pH reading of 7.0 is neutral for soils measured in water paste.

Phenotype: All characteristics of a plant, morphological, anatomical, and physiological, as determined by the interaction between genotype and the environment.

Plant Moisture Stress (PMS): Measure of plant water status, equal to the absolute value of plant water potential PMS is an integrated index of the current moisture status of a plant and is usually influenced by soil moisture status and evaporative influenced by soil moisture status and evaporative demand of the atmosphere.

Plant Water Potential: Current water status of a plant consists of two components, turgor potential and osmotic potential and is usually measured in negative pressure units (– bars)

Plant Per Cent: The percentage by number, of seeds which develop into plantable plants, seedlings or transplants, at the end of a given period usually the end of the nursery period at the time of field planting.

POD: A superior, one-celled, one-or many-seeded dehiscent fruits of two values.

Pore Space: Total space not occupied by soil particles in a built volume of soil.

Porosity: Volume of total soil bulk not occupied by solid particles expressed as a % age. Percent porosity equals the volume of pores divided by total soil volume.

Post Emergence: Time period after the crop plants or weeds emerge through the soil surface.

Prechilling: Cold moist treatment applied to seeds prior t o sowing in the soil or germination in the laboratory, and designed to hasten after ripening or to overcome dormancy.

Procuring: The deliberate storage and slow air drying under shade of fruits and contained seeds in order to render them more suitable for subsequent operations, e.g., kiln drying, extraction and storage.

Preemergence: Time period before crop, plants or weeds emerge through the soil surface.

Replanting Treatment: The application of herbicides/or fertilizers before a crop is planted.

Profile (Soil): A vertical section of soil extending through all of its horizons and into the parents material.

Pure Seed: That component of a seed lot which usually consists of the designated species. Excludes the seeds of other species, wings of coniferous seeds, seeds of coniferous or leguminous species with seed coats entirely removed, broken seeds particles less than half the original size other matter such as stones, twinges and the leaves.

Purity: Proportion of clean, intact seed of the designated species in a seed lot, generally expressed as a percentage by weight.

Quiescence (Seedling): Period of physiological inactivity preceding true dormancy when the plants still can grow, if, environmental conditions are suitable.

Quiescent: Inactive, resting. Applicable to non-dormant seeds during the internal between maturation on the parent tree and the onset of germination.

Radiation Frost: Frost occurring when large amounts of heat in the soil and plants are dissipated into the atmosphere, allowing temperatures near the ground to reach the freezing point. Generally occurs on cloudless nights.

Radical: It is a portion of the axis of an embryo from which the primary root develops.

Replication: Repetition of a treatment in an experiment.

Residual Sodium Carbonate (RSC): Irrigation water quality index that measures the difference between the sum of calcium and magnesium ions and the sum of the carbonate and bicarbonate ions. RSC values are used to calculate the effect of salts on the soil permeability.

Ripping: Cultural practice used to ameliorate compacted sub-soil by pulling shanks through the soil at a depth of 40 to 80 cm. Usually the shanks are then pulled at right angles to the first pass to produce a grid pattern.

Rolling: Cultural practice used sowing to ensure good contact between seeds and soil particles. A cylindrical roller is passed over the land to firm the soil without causing a great deal of compaction.

Root Culturing: The nursery cultural practices designed to modify the seedling root growth, e.g., undercutting, wrenching, etc.

Root Growth Capacity: It is also called root-growth potential. Physiological capability of a plant for producing new roots under optimal environmental conditions for root growth. High root-growth capacity is thought to be one of the best indicators of seedling field survival.

Salt: Water soluble chemical compound that either is found naturally in the soil or is applied as fertilizer. In soil solutions, salts are generally dissociated into cations (e.g., calcium, sodium, potassium, and anions (e.g. sulphates, nitrates, bicarbonates,). High salt levels are detrimental to plant growth. Conifer seedlings are essentially sensitive to the salts.

Sand: Soil particle between 0.05 and 2.00 mm in diameter; soil texture class characterized by a predominance of sand particles.

Scarification: Disruption of seed coats, usually by mechanical abrasion or by brief chemical treatment in a strong acid, to increase their permeability.

Seed: A fertilized and mature ovule which contains an embryo nutritive tissue and is enclosed in protective layers of tissue (Seed coat).

Seedbed: Elongated strip of prepared soil in which seeds in which seeds are sown and seedlings raised.

Seedbed Density: Number of seedlings growing in a seedbed, expressed relative to area (e.g., number per square meter or foot) or lineal measures (e.g., number per lineal meter or foot).

Seedling: Young free propagated from the seed.

Seedling Quality: Potential of a seedling to survive and grow successfully after out planting.

Seed Lot: A specified quantity of seed of reasonably uniform origin and quality.

Seed Protectant: Pesticide applied to seed before planting to protect seeds and new seedlings from diseases and insect-pests.

Seed Quality: Refers to the purity, germination capacity or vigour of a seed lot.

Seed Year: In respect of any species mainly the trees of irregular or infrequent seed production a year in which it produces, either an individual or a crop, an adequate amount of seed. Many periodic seeders produce heavy (bumper) seed crops during their seed years.

Seed Zone: Area of similar environment conditions. Plant originating from the same seed zone are believed to be similarly adapted to the environment.

Standard Seed: For a given species, seed with 'average' quality characteristics based on previous experience with that species. Used with particular reference to the seedling recovering figures to be expected from 1 kg of standard seed and its relationship to actual seedling recovery from a particular seed lot.

Stratification: Practice of burying seeds in moist medium, often in alternate layers to overcome, often in alternate layers to overcome dormancy; commonly applied to any technique which keeps seeds in a cold and moist environment.

Submitted Sample: The sample of seed submitted to a seed testing station.

Substrate: Under layer of soil, sand and other matter. The matter on which a fungus or germinating seedling grows.

Silt: Soil particle between 0.05 and 0.002 inch-meter soil textural class characterized by a predominance of silt particles.

Soil Horizon: Layer of soil approximately parallel to the land surface and differing from adjacent genetically related layers in physical chemical and biological properties or characteristics such as colour, structure, texture, consistency, amount of organic matter and degree of acidity or alkalinity.

Soil Test: Chemical or physical analysis of a soil to determine texture acidity total salt concentration or concentration of nutrient elements.

Soil Water Potential: The amount of work that a plant must exert to absorb water from the soil generally expressed as in the units of negative pressure (–bars).

Sowing: Process of placing the seeds in the seedbed at a specific depth and density.

Specific Gravity: Ratio of weight of a volume of matter to weight of an equal volume of water at a specific temperature.

Stock Type: Seedling classification generally by age and location in the nursery.

Surfactant: Chemical agents, e.g., spreaders, detergents, wetting agents added to pesticides to make mixing easier and to assist application of a solution and adherence to the treated surface.

Systemic Herbicide: The herbicide which is absorbed by and then distributed within a plant as opposed to one which functions only on contact with the plants surface.

Table Pruning: Pruning of seedling roots at the time of grading and packing.

Top Dressing: Application of chemical or fertilizer after a crop has been established.

Transpiration: Process of water movement through a plant to the atmosphere as a result of evaporation of water from the leaves.

Transplant Stock: Reduced growth rate of a young tree after it has been transplanted or out planted.

Tree Percent: Number of seedlings irrespective of size or form in a nursery bed at lifting compared to the number of viable seed sown, expressed as a percentage.

Tumbling: The operation by which cones or fruits are placed in a rotating drum, which rolls and tosses them around to induce the release of the contained seeds.

Turgor Potential: One of the main components of plant, water potential reflects a positive force extended inward by the cell wall.

Undercutting: Root pruning in the nursery bed using a sharp blade drawn parallel to the soil surface at a regulated depth to stimulate root growth and fibrosis.

Viability: Ability of a seed to germinate and grow under a given set of condition usually estimated by germination per cent or other tests.

Viability Per Cent: At the end of a germination test, the total number of germinated seeds plus ungerminated but sound seeds (on cutting test) expressed as a percentage of the number of seeds sown.

Viable Seed: A seed that can germinate under favorable condition, provide that any dormany that may be present it just removed.

Vigrou: Those seed properties which determine the potential for rapid, uniform emergence and development of normal seedlings under a wide range of field conditions.

Water Content: Index of soil moisture status calculated as the amount of water lost from the soil upon drying to constant weight at 105°C generally expressed as the weight of water per unit weight of dry soil.

Wettable Powder (WP): Powder formulation of a pesticide which contains a welting agent so that it will readily from a suspension in water.

Wind Frost: Frost occurring when winds in excess of 4mph from cold regions displace warmer air. Wind frost can occur day or night and are not necessarily dependent on topography.

Winter Desiccation: Type of foliage injury which occur on warm days when the ground is usually frozen, actually a type of physiological drought caused by excessive transpiration when the frozen soils prohibit water absorption.

Winter Scald: Type of cold injury to tree bark. The bark is warmed above freezing by the winter sum, during the day time, even though at temperature then refreezes after then refreezes after sunset.

Working Sample: A reduced seed sample taken sample meet specific ceterion regardless of form expressed as a percentage. These seedlings may have multiple tops or damage from diseases, insdepests or some other agents characteristics that may make then unacceptable for shipping purposes.

SOURCES OF INFORMATION

The following sources have been used in compiling this selected glossary:

Belmont BL 1983 the new pesticide user's guide. Reston Publishing Co., Inc. Reston, Virginia, 452 p.

Clearly BD RD Greaves and R.K. Hermann (eds.), 1980 Regenerating Oregon's Forest, Oregon State. Extension Service, Corvaliis, Oregon 286 p.

Haman II WD Holley and KL Goldsberry 1978 Green House Management. Springer-Verlag. Berlin. Heidelberg, New York, 530 p.

SSSA 1965 Soil Science Society of America, Glossary of soil science terms, SSSA Proc. 29(3)L 330-351.

Williams RD and SH Hanks 1976 Hardwood nursery man's guide, USDA, Forest Service, Washington, DC. Agric. Handbook No. 473., 78 p.

REFERENCES

Aldhous JR 1972 Nursery Practice Forestry Commission Bull. 43. London: Her Majesty's Stationary Office. 48 p.

Alison FE 1965 Decomposition of wood and bark sawdust in soil, nitrogen requirements and effects on plants, U.S. depth. Agric, Washington, D.C., Tech. Bull. 1332. 29p.

AOSA 1981 Association of official seed analysts rules for testing seeds. Jour, of seeds technology 6(2): 1-126.

Barres H 1964 Rooting media for growing pine seedlings hydroponic culture. Res, Note. ITF-2, Rio Pieclras, PR; U.S. Department of Agriculture, Forest Service, Institute of Tropical Forestry, 4 p.

Baur CN 1960 Nursery spacing and grading of slash pine seedlings. Division of forest management. Forestry Common of New South Wales, Research Note 5. 7 p.

BCFT 1953 British Common Wealth Forest Terminology, Part I. Empire Forestry Association, London (U.K.)

Benzian B 1965 Experiments on nutrition problems in forest nurseries. Her Majesty's Stationary Office, London Forestry Commun Bull, 37, 251p.

Bin Haji Ahmad Darus 1980 effect of fruits size upon germination of teak (Tectona grandis). The Malaysian Forester 43(3): 396-397

Bin Mohamad Aminuddin and Bin brahim Zakaria 1980 Grading of Gmelina arberea (Yemane) fruits by colour. The Malaysian Forestor. 43.(3): 337-339.

Bonner FT 1974 Seed Testing. In: Seeds of Wordy Plants in the United States. Agriculture Handbook No. 450, For. Service, USDA, Washington, D.C.

Burschel Peter Moll Wolfgang and Soto Manuel 1973 Siltuatioon Y Marejo de los viveros chilemnos de pino insigne (Pinus radiate D. Don). Publicaciones Cientificers. Valdivia, Chile: Facultad de Ingenieria Forestall, Universidad Astraddle Chile 26 p. (Original not Seen.)

CFASIC 1980 California Fertilizer Association Soil Improvement Committee Western Fertilizer Handbook. The Inter State Printers and Publishers, Inc. Danville, Illinois, 269 p.

Carter AS 1961 In Testing the sample is all important, In: Seed year Agr. (USDA). Pp, 414-416.

Cordell Charles E and Filer T.H. Jr. 1984 Integrated Pest Management. In: Lantz, Clark W., ed. and comp. Southern Pine Nursery Handbook, Atlanta, GA; US Department of Agriculture, Forest Service, Southern Region, 13 : 13-17.

Davey CB and HH Krause 1980 Functions and Maintenance of organic matter in Forest Nursery Soils Pages 130-165. In: Proc. North American Forest Tree Nursery Soils Workshop (L.P. Abrahamson and D.H. Bicelhaupt (eds.) State University, New Yord. Coll. Environ., Sci, and Forestry. Syracause. Davidson H and Mecklenburg and culture. Englewood Chiffs. NJ: Pretice-Hall). 450 p.

Epstein E 1973 Mineral Nutrition of Plants: Principles and Perspectives New York: John Wiley and Sons. 412 p.

Etter HM 1971 Nitrogen and Phosphorus requirements during the early growth of white spruce seedlings. Can J. Plant Sci. 5(6): 63.

Evans J 1982 Plantation Forestry in the tropics, Claremden Press. Oxford; 472 pp.

Weart AJ 1908 Proceedings of the royal society of Victoria Melbourne. 21(1).

Fair lamb J and Davidson J 1966 Germination of teak. Seeds-Preliminary evidence of a chemical of chemical germination inhibitor, In: "Seed Problems". Proc. Second international Symposium on physiology of seeds germination, IURFO, Ferji, Japan, Oct., 1976.

Fishwick RW Irrigated Nursery practice of animal instruction, Kaduna, Northern Nigeria: ministry of animal and forest Resources; 19 p.

Galloway Glen 1981 Guia para of el establicimient to Y manteinmiento de un peyeno vireo forestall. Informe de investigation Forestall 72. Carton de Columbia, S.A ; Cali Columbia; 11 p. (original not seen).

Gour AY and Barney CW 1976 Forest tree planting in arid zones (2nd edi.). Ronald Press, New York.

Greaves A 1981 Gmelina arborea-a review of the literature. Review article for. Abst. Vol. 42 (No. 6.) Common wealth for. Bur. Oxford.

Groome JS Less HMN and wig LT 1975 A summary of information on pteroearpus angolensis. Leading Article in For. Abst Vol. 18, 3-8, 153-162. Comm. For. Bureaus Oxford.

Gupta BN and Kumar Adarsh 1977 Interrelated effects of temperature and moisture on seeds germination of dendroclalamus strictus neis. The India Forester. 103(3): 212-19.

Geary TF Avila Cortes G Hadley HH 1971 1971 germination and growth of pinus caribaeu and captan. Turrialbe 21(3): 336-42.

Heit CE 1967 b Propagation from seed part 10: storage methods for conifer seeds. Am. Nurseryman 126(8).

Holmes GD and Burze wiez G 1958 the storage of seeds of temperature forest tree species For. Abstr 19: 313-22, 455-76.

Howcroft NHS 1973 A direct sowing technique for pinus. Pupua,New Guinea, Department of forest ap.

ISTA 1976 International seeds testing association international rules for seeds testing seed science and technology, 4: 4-177.

Jackson JK 1974 Nursery Techniques In the savama region of Nigeria Ministry of Agriculture and Natural Resources. Res. Paper 32. Ibadan, Federal Republic of Nigeria, Federal Department of Forest Research. Sp.

Justice OL 1972 Essential of seed testing. In: Seed Biology, Vol. 3 (ed. T.T. Kozlowski) Academic Press New York and London. 301-70

Kamra SK 1973 X-ray Radiography of teak seed (tectona grandis L.). In: "Seed processing". Proc. Symposium IUFRO WKg group on seed problems Bergen vol. 1. Paper 9, 14 p.

Kemp RH 1976 seed procurement for species and provenance research. In: A manual on species and provenance research with particular reference to the tropics. (ed. J. Burlug and P.J. wood). Trop. For. Paper No. 10. Common Wealth, For. Inst. Oxford.

Khanna LS 1984 Principles and practice of silviculture Khanna Bandhu, Tilak Road, Dehradun (Uttarakhand), India, 473 p.

Kittredge Joseph 1973 Forest Influences: the effects of wood vegetation on climate, water, and soil with applications to the conservation of water and the control of floods and erosion. New York: Dover, 394 p.

Leanard dave 1980 soils, crops and fertilizers use or a what, how and why guide 3rd, ed. Information collection and exchange reprint RS. Washington, D.C. Peace crops. 162 p.

Liegel LH and Venator CR 1987 A technical guide for forest nursery management in the Caribbean and Latin America, U.S.D.A., Forest service New Orleans, Louisiana, 156 p.

Lowe RG 1967 A tropical medium for tree species tech, note 37, Ibadan, Federal Republic of Nigeria; Department of Forest Research, sp.

Magini E 1962 Forest Seed Handling, Equipment, and Procedures, II. Seed treatments, storage, testing and transport, Unasyliva 16(1) 20-35.

McDonald SE irrigation in forest-tree nurseries: monitoring and effects on seedling growth, In Duryea, Mary L and Land is Thomas D.eds. Forest Nursery manual: Production of bare-root seedlings. The Hague the Netherlands: Martinis NIjho off/Dr W. Junk Pub; Covallis, OR; Forest Research Laboratory, Oregon state university, p. 107-121.

Minko G and Craig FG 1976 Radiata Pine nursery research in North-Eastern Victoria, Forest commission Bull, 23, Melbourne, Australia, 24 p.

Matias AR Betancourt A Zayas A pena A and Rivo R 1973 Forest seeds in Cuba, In: "Seed Processing". Proc. Symposium IURFO Wkg group on seed problems Bergen. Vol. II, Paper 31.

Marrero Jose (1947 the proper depth and kind of covering for seeds of several tropical hard woods. Caribbean Forester 8(3): 213-36.

Marrero Jose 1948 Forest Planting in the Caribbean National Forest, Part experience as a guide for the future, Caribbean Forester 9: 95-213.

Marrero Jose 1961 Shapgnum moss as a medium for rooting pine seedlings. Trop. for. Notes 7. Rio Piedras, PR: US Department of Agriculture. Forest Service. Institute of Tropical Forestry 2p.

Marrero Jose 1962 Practices usaders enlos viveros de pinos de Puerto Rico. Caribbean forester 23(2): 87-99.

Marrero Jose 1965 Effect of Plastic mulch on weed growth and early height growth of Honduras pine. Res note ITF-4. RIUO POiedras, PR: US Department of agriculture. Forest Service. Institute of Tropical forestry 5p.

Monteith DB Yharra Coronodo R 1969 Survival of bare-rooted Honduras Pine in Puerto Riuco. Turrialha 19(21): 291-92.

Paul Derek K 1972 A handbook of nursery practice for pinus caribaea var. hondurensis and other conifers in west Malaysia. United National Development Programme. Food and Agriculture organization of the united nation; Fo : SF M.AL 12. Woring paper 19 Kaula Lumper, Malaysia, 139 p.

Peace crops 1982 Forestry training manual Washington DC. Inter American Region Office of Programme Development. Forestry Natural Resource Sector Peace Corps 390 p.

Perera WRH 1973 Seed for Sri Lanka's reforestation programme, paper No. 23, In: Proc. International Symposium on seed processing IUFRO working purty: S2.01.06 Bergen Norway, vol. II. 14 9.

Pattanuth PG 1982 Personal Communication.

Robbins AMJ Irrimecu MI and Calcderon R 1981 Recoleccion de Semillas Forestalls Pub. Mise. No. 2. Esculea Nacinal de Ciencias Forestales Siguatepegue, Honduras (original not seen).

Roberts EH 1973 Predicting the storage life of seeds. Seed Sci. and Technol. I. 499-514.

Reddy SR 2004 principles of agronomy Kalyan Publishers, Ludhiana (Punjab), India. 543 p.

Reddy MAR and Mishra BM 1970 Fungicidal soil treatments to control damping-off diseases in pines, Ind. For.

Sagwal SS 1978 Genetic variation and selection of plus trees in chil (inus roxburhii sargent) of Himachal Pradesh M.Sc. Forestry Thesis un-publishing submitted to H.P. Krishi Vishivaridyalaya, Palampur (H.P.), India.

Seeber G and Agpaoa A 1976 Forest Tree seeds. In Manual of reforestation and erosion control from the Philippines 473-535 German Agency for technical co-operation Esch. Born.

Sheikh Mahmood Iqbal 1979 Tree seeds respond to acid scarification Pakistan Journal of Forestry 29(4): 253-54.

Stead JW 1979 Exploration collection and evaluation of cordial alliodora (R & P) Oken. For. Genetic resources information No. 9. FAO. Rome.

Stein WI slabaugh PE and Plummer AP 1974 Harvesting, processing and storage of fruits and seeds, In: seeds of woody plants in the united states agroculture handbook No. 450. For. Service, USDA Washington, D.C.

Sutter Land JR 1979 the pathogenic fungus caloscypha fulgens. In stored conifer seeds in British Columbia and relations of tis incidence to ground and squired cache collected cones con. J. For. Res. 9: 129-132.

Solberg Karl H 1978 Direct sowing contra transplanting in the nursery Pinus Caribeae and P. Oocarpa, Tanzania EAC/NORAD Lowland afforestation Project Tech. Rep. 7. As. Norway: Department of silviculture. Agricultural university of Norway (NLH): 8p.

Solberg Karl H 1978 Miscellaneous nursery experiments, pinus caribaea and P. P oocarpa-tanzinia EAC/NORAD Lowland afforesation project Res. 8. Aas, Nowray: Department of silviculture. Agricultural university of Norway (NLH). 33p.

Pharis RP and Kramer PI 1964 the effect or nitrogen and drought on loblolly pine seedlings forest sci, 10: 143-150.

Rogers TH and JE Giddens 1957 green manure and cover corps pages 252-257. In: Soil. The yearbook of agriculture 1957 U.S. dept. agric. Washington, D.C.

Tanaka Y 1984 Assuring seed quality for seedling production cone collection and seed processing, testing storage, and stratification In: Duryea marys and landis Thomas D., eds. Fort nursery manual: Production of Bare-root seedlings. The hague the Netherlands: Martinus Neijoff/Dr. W. Jank Pub: Corvallis, OR: forest, Research laboratory Oregon state university 27-39.

Theron JM 1972 Nursery Research in Malawi 1964-1971. Research Record 56. Zomba: Malawi forest research institute 43 p.

Timmis R 1974 effect of nutrient stress on growth bud set and hardiness in Douglas fir seedlings, pages 187-193. In: Proc. North-American American containerized forest tree seedling symp. (R.W. Tinus, WI stain and WE Balmer. eds). Great Plains Agric. Council publ.

Tiren L 1947 On en snabb `method of gro-barhet=sbestimning are tall-och grandro. Medd from statens skigs forskniings institect. Band 37, No. 5.

Troup RS 1921 The silviculture of India Tress. 3 Volumes, Clarendon Press, Oxford.

Turnbull JW 1975 Seed extraction and cleaning: Report on FAO/DANIDA training course on forest seed collection and handling. Vol. 2. FAO, Rome.

USDA 1974 united states department of agriculture forest service. Seds of woody plants in the united states Agric. Handbook no. 450. Washington, DC. 4s Department of agriculture forest service 883 p.

Tisdale Samuel L and Welson Werner L 1975 Soil fertility and fertilizers, New York: Mac Millian 694 p.

Venator CR 1972 effect of gibberelic acid germination of low-vigour Honduras pine seeds. Forest science 18(4): 331.

Venator CR 1973 The relationship between seedling height and date of germination of pinus caribaea var.hondurensis. turrialba. 23(4): 473-474.

Venator CR and zambrana JA 1975 extraction and germination of kadam seds. Res.note ITF-14A. Rio Piedras, PR; U.S. department of agriculture. Forest service, institute of tropical forestry, 3p.

Voorhoeve AG weelderen AWH 1968 nursery practice of Pinus caribaea, Paramaribo survinam forest service. Research Branch 20p. Wads worth frank H and marreso Jese 1953 The signififance to Puerto Rico of comparibia paulista experience with eucalypts. Caribbean forester. 14(1-2): 65-78.

Weber Freed R 1977 reforestation in arid lands peace crops/vita publications manual series 37E. Washington, DC. Government printing office, 248 P.

Weidelt HJ comp 1975 manual of reforestation and erosion control the Philippines, Eaclborn, Germany; Agency for technical cooperation Ltd. (472). 569p.

Wilde SA 1959 forest soils. The Ronald Press Co., New York 537 p.

Wunder WG 1966 The handling of seed in sudan forestry, Pamphlet No. 19. Sundan Forests Department and UNDP Forestry Research and Education Project forest Res. Inst., Soba, Khartoum.

Van den driessche R 1984 soil fertility in forests nurseries. In: Duryea Mary L and Landis Thomas D. eds Forest Nursery manual; production of bare root seedlings. The Hague the nether lands: martinus nigjoff/Dr. W. Junk pub; Corvallis or forest research laboratory, Oregon state university. 75-80.

Youngberg CT soil and tissue analysis tools for maintaining soil fertility. In: durgea mary L and landings Thomas D., eds. Forest nursery manual: production of bare root seedlings. The Hague the Netherlands; Martinus, Nijhoff/Dr. W. Junk, Pub. Corvaillis, or: forest research laboratory, Oregon state university, 75-80.

Simposon JA nursery nutrition studies with slash pine at beerburiu: Queensland Res. Paper No. 9. Brishbane, Queensland's Australia Department of Forestry. 32 p.

Simpson JA use of inorganic fertilizers and cover crops in exotic pine nurseries of Southern Queensland Australia, In: South David B., ed., proceedings, international symposium on nursery management practices for the southern pines, 1985 August 4-9; Montgomery. AI. Aubrun AI; Schol of forestry Alabama Agricultural experiment station, auburn university; 203-212.

Appendix

Appendix–I

Potential Nursery Site

1. Name of Area ____________ Date ____________
2. Location ____________
3. Soil Survey Type ____________
4. Top Soil
    1. Textural Assessment ____________
    2. Depth ____________
    3. pH ____________
5. Sub-Soil
    1. Textural Assessment ____________
    2. Depth ____________
    3. pH ____________
6. Depth of Water Table ____________
7. Drainage ____________
8. Water Supply
    1. Adjacent Creek ____________ ____________
    2. Well ____________
    3. Other ____________
    4. Rise or Fall (Feet) to Supply ____________
    5. pH ____________
9. Topography
    1. (a) Level ____________
       (b) Rolling ____________
       (c) Irregular ____________
    2. (a) Slope ____________ to ____________ Degree
    3. Unsuitable (i.e., more than 6° or very irregular) ____________

10. Cover
    1. Main Tree Species ____________________
    2. Common Members of Ground Flora ____________________
11. Climatic Data (Use nearest station in department of transport table)
    1. Total Annual Precipitation ____________________
    2. Spring Precipitation March ____________________
       April ____________________ May ____________________
    3. Number of Frost-Free Months ____________
    4. Highest Temperature (and month recorded) Last Year ____________
12. Approximate Acreage ____________________
13. Access ____________________ Ownership

[A Nursery-site checklist such as this one is recommended for the use by the site-selection team (adopted from Van den driessche R. 1969. Forest Nursery Handbook B.C. Forest Service. Victoria Res. Note. 48, 44 p.).]

APPENDIX – II

List of Equipments of Seed Collection and Site Information

A. Seed Collection:

1. Seed containers (field)
2. Sacks and bergs (may be re-used)
3. Seed containers (Despatch)
4. Cotton bags, Convas Sacks (Desptach with seed)
5. Tree markers, e.g., plastic tupe.
6. Climbing equipment:
   - 6.1 Foot spurs
   - 6.2 Ladders
   - 6.3 Safety belts
   - 6.4 Safety ropes
   - 6.5 Safety helmets
   - 6.6 Tool lines
7. Seed cutters, e.g., are hooks, cone rakes, pruning shears, hand secateurs.
8. Plastic sheeting (heavy gauge) for,
   - 8.1 Protection when storing fruits/seeds;
   - 8.2 Extracting seeds.
9. Binoculars: For studying
   - 9.1 Tree crowns
   - 9.2. Fruit development.

10. Insecticidal and fungicidal powders for seed protection.
11. Others: Axe, Saw, machete, knife, ripe, string, labels, marking pens.

B. Site Description

1. Notebook, description forms.
2. Maps: Outline copies also for plotting.
3. Compass and altimeter.
4. Meteorological equipment: Hygrometer, maximum/minimum thermometer.
5. Soil survey equipment: Augar, colour charts, pH test kit.
6. Tree measuring equipment altimeter diameter tapes, bark gauge, etc.
7. Camera and equipment (if, required)
8. Spude.

APPENDIX – III (A)

Area: Conversions

1. Abbreviations

English	**Metric**
Square inches (in^2)	square millimeters (mm^2)
Square feet (ft^2)	square centimeters (cm^2)
Square yard (Yd^2)	square meters (m^2)
Square miles (mL^2)	square kilometers (km^2)

2. Conversions

English to English	**Metric to Metric**
1 ft^2 = 144 in^2	1 cm^2 = 100 mm^2
1 Yd^2 = 9ft^2	1 m^2 = 10,000 cm^2
1 ac = 4840 Yd^2	1 ha = 10,000 m^2
1 mi^2 = 640 ac	1 km^2 = 100 ha
1 mi^2 = 27,880,000 ft^2.	

English to Metric	**Metric to English**
1 in^2 = 6.452 cm^2	1 cm^2 = 0.1550 in^2
1 ft^2 = 0.09290 m^2	1 m^2 = 10.769 ft^2
1 yd^2 = 0.8361 m^2	1 m^2 = 1.196 yd^2
1 ac = 0.4047 ha	1 ha = 2.471 ac
1 mi^2 = 259.0 ha	1 ha = 0.003861 mi^2.

APPENDIX – III (B)

Length

Abbreviations

English Metric

inches (in) millimeter (mm)

feet (ft) centimeter (cm)

yard (yd) meter (m)

mile (mi) kilometer (km)

Conversion

English to English	**Metric to Metric**
1 ft = 12 in	1 cm = 10 mm
1 yd = 3 ft	1 m = 100 cm
1 m = 1.760 yd	1 km = 1,000 m
1 m = 5.280 ft	
English to Metric	**Metric to English**
1 in = 2.54 cm	1 cm = 0.3937 in
1 ft = 0.3048 m	1 m = 3.281 ft
1 yd = 0.9144 m	1 m = 1.094 yd
1 m = 1.609 km	1 km = 0.6214 m

APPENDIX – IV

Species: Botanial Name: ______________________________

Common Name: ______________________________

Date of Collection: Month ______________ Year ________

Location of the Seed-Producing Stand

Region: ______________ District: ______________

Locality ______________________________

Latitude: ______________ Longitude: ______________

Altitude: ______________ Aspect: ______________

Inclination: ______________________________

Association: ______________________________

Geological Condition and type of soils: ______________

Remarks: ______________________________

____

____

Description of the Seed-Producing Stand

How originating:

Present Age:

Present treatment applied:

Seed-Bearers:

(i) Diameter;

(ii) Height:

Stem form:

Nature of Branches:

(i) Size

(ii) Inclination

Remarks: ____________________

____

____

Certificate: The above particulars are certified correct by the competent forestry officer.

__________SIGNATURE__________

The above particulars have been submitted to:

APPENDIX – V

1. Insecticides

1.1 Malation: Widely available, sometimes odorous; useful to control aphids, scales, mealy bugs, and Chewing insect.s

1.2 Metaldehyde: Useful for control of snails and slugs.

1.3 Nicotine Sulphate: An old, reputable, and effective control for aphids and other sucking insects.

1.4 Ethion: Useful in controlling leaf-hoppers, mites and thrips.

1.5 Rotenone: A plant material used both as a contact and stomach poison; it has no toxicity to man.

1.6 Carboryl: A broad spectrum insecticide, it is very toxic to bee populations and should be avoided where production lives are located.

2. Miticides

2.1 Supracide 2E: Long shelf-life if stored in coal, dry place.

2.2 Ovex: Extremely effective but use requires special care, read label instructions very carefully.

2.3 Chcorobenzilate: Used with good results.

3. **Fungicides**

   3.1. Methyl Thiophanate: Effective for control of soil fungi, used as a soil drench.

   3.2 Bordeux Mixture: Probably the most practical fungicide available. Use immediately because effectiveness is lost, if, the mixture stands too long. Adding a little sop to the final solution makes it stick and spread better

   3.3 Formaldehyde: Also known as 'formalin'. Can be used on bare ground before planting, it is toxic to plants jet excellent as a pre-planting drench to control soil fungi, its fumes are very toxic to people.

   3.4 Lime Sulphur: Used in controlling spot diseases, i.e., leaf spot fungi.

4. **Soil fumigants**

   4.1 Ethylene dibromide: An effective soil fumigant that controls soil insets and nemnutodes. Soil treated with material cannot be planted for at least with material cannot be planted for at least three (3) weeks because it is very toxic to the plants. It is also very toxic to humans. It is sold under a number of commercial names.

   4.2 Nicotine: Sold under several trade names, it is safe to use and effective for soil insects.

5. **Spreaders:** Substances to increase the adherence of sprays to plants. A few shavings of ordinary soap or a small amount of detergent will serve the purpose.

6. **Thiram:** A Wettable powder used as a slurry to coat seeds for protection against damping-off, seed decay, seedling blight and rodents. It is usually suggested for conifer seeds.

7. **Pre-planting Bed Treatment:** Applied to nursery beds before planting to control dumping-off; it is worked into the top 8 to 10 cm of soil.

Source: Liegel, L.H. and Vinator, C.R. (197). A Technical Guide for Forest Nursery Management in the Caribean and Latin America, General Technical Report 50-67, USDA (USA).